S0-EGE-273

LIBRARY SYSTEMS ANALYSIS GUIDELINES

LIBRARY SYSTEMS ANALYSIS GUIDELINES

EDWARD A. CHAPMAN,

Director of Libraries,
Rensselaer Polytechnic Institute,
Troy, New York

PAUL L. ST. PIERRE,

Consultant,
Library Community for Data Processing Education
Albany, New York

JOHN LUBANS, JR.,

Coordinating Librarian,
Reader Services Division,
Rensselaer Libraries,
Troy, New York

Wiley-Interscience, a Division of John Wiley & Sons Inc.

New York · London · Sydney · Toronto

Library of Congress Catalogue Card Number: 75-109391

SBN 471 14610 2

Printed in the United States of America

10 9 8 7 6 5 4 3 2

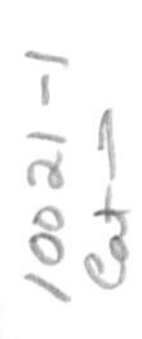

PREFACE

This book aims primarily to provide guidelines for library administrators and library systems analysts in analyzing and evaluating existing operating systems and in designing new or improved ones. In addition, we have found the material to be very useful in staff training programs conducted to ensure knowledgeable staff participation and cooperation in a systems study. This guide is also adaptable for introducing library school students to the concepts of systems study in the library.

The guidelines set forth here are based on experience in applying scientific management concepts and techniques to operations concerned with processing and manipulating records inherent in the operation of a library. Application of these techniques is based on the "total systems" concept, which we believe to be the major premise of an effective systems study and the logical course for the design of new procedures. Details are furnished for the data processing operations carried on in the acquisitions system, the cataloging system, the serials control system, and the circulation control system.

In library as well as business management literature, the terms "data processing" and "systems study," "systems analysis," or "systems engineering" usually are equated to the use of computer-based systems. It is true that a systems study in minute detail is a prerequisite to a well-designed electronic data processing system. However, the result of a systems study need not be a computer system. The organization of information handling operations can and should be a matter of study, whether operations are to be computer based or existing operations are to be revised. It is, we think, the study for improvement of the data processing procedures that is fundamental to the library's role in actively serving the information needs of its community.

This guidebook began in the thought that others might profit from the experience of the Rensselaer libraries in the study and application of scientific management principles to library operations. Further encouragement was received from favorable staff reaction to extensive and rigorous training programs in which many of the area librarians faithfully participated. An outline manual was prepared for the guidance of the participants, and this manual was evolved into the present book.

Encouragement to develop the manual into a book was given by the Upper New York State Chapter of the Special Libraries Association which sponsored a two-day institute on "Systems Study" in September 1966. The Connecticut and Montreal Chapters joined in this institute, which received an enthusiastic response. Recognition of the potential value of the systems study concept to libraries by many librarians—academic, public, and special—and by a sampling of library school directors and faculty strengthened the determination to proceed.

We also acknowledge the encouragement received from Stephen R. Salmon, first President of the Information Science and Automation Division of the American Library Association, and his executive committee; we participated in the Division's Preconference Institute in San Francisco (June 1967) where the topic was presented to the Conference. Steady requests from librarians here and abroad for the preliminary manual played a significant part in deciding to publish a fuller treatment of the subject.

The first draft of this guide was used as the teaching manual in a three-week program during the summer of 1968. This institute was supported by a grant from the U. S. Office of Education under the provisions of Title IIB of the Higher Education Act of 1965.

The encouragement and assistance of Verner W. Clapp, Joseph Becker, and Robert M. Hayes, are gratefully acknowledged. Verner, in the year of his retirement (1967), arranged for the necessary financial support of the project through the Council on Library Resources, and barked, "Get it out, the sooner the better"; Joe and Bob review the work.

The portion of the chapter on reporting the findings of the systems study dealing with the principles of report writing was developed in consultation with Professor Robert A. Sencer of the graduate technical writing curriculum at Rensselaer. His contribution cannot be overestimated, since many a well-conceived program may flounder because of a poor written presentation.

Many others helped during the development of this book. The patience, understanding, and aid of the Rensselaer library staff must be cited; without its cooperative participation this guide would not have been written. As in building a new library, so with this work, undesirable features undoubtedly will reveal themselves, however good the plans were thought to be. Criticisms and suggestions are sincerely sought.

EDWARD A. CHAPMAN
PAUL L. ST. PIERRE
JOHN LUBANS, JR.

Troy, N. Y.
New York, N. Y.
November 1969

CONTENTS

1

SYSTEMS IN A LIBRARY

2

PLANNING AND CONDUCTING THE SYSTEMS STUDY

9

SYSTEMS DESIGN—COMPUTER-BASED ACQUISITIONS SYSTEM

10

SYSTEMS DESIGN—COMPUTER-BASED SERIALS SYSTEM

INTRODUCTION

The Committee on Research Libraries of the American Council of Learned Societies reports that "research" libraries are in serious difficulty arising from shortages of space, staff, and funds in the face of greatly increased demand for services,[1] resulting in increased complexities in the obligations of library management. The Committee offers "no panaceas for the problems of the research libraries" but has "no doubt that they can be overcome by a carefully designed combination including a substantial admixture of research and development." [2] It is a commonly observable fact that all types of libraries are faced with approximately the same problems to which may be added the latter-day one of book funds increased by incentive federal grants that, in many instances, create the problem of increased personnel funds.

From the managerial point of view the chaotic operational conditions caused by a combination of these problems become subject to systematic scrutiny. The solution may lie in simply adjusting present procedures or in a combination of procedural adjustments calling for the application of technological aids including the computer. To exercise either option management would wish to base a decision on a systematic analysis because if the analysis dictates the computer-assisted solution, management may need to commit substantial additional operating funds and capital expense for input equipment. Only a rigorous analysis will make certain that, in the context of the interaction among all the library systems and within the community or organization for which the library exists, the system contains all of the elements or steps it needs to operate at maximum effectiveness.

As has been demonstrated many times in the past eight to ten years,

[1] *On Research Libraries; Statement and Recommendations of the Committee on Research Libraries of the American Council of Learned Societies; Submitted to the National Advisory Commission on Libraries; November 1967,* MIT Press, Cambridge, Mass., 1969, p. ix.

[2] *Ibid.,* p. ix.

the period of "discovery" of the computer as a library tool, the decision to automate is to be approached with extreme caution and thorough deliberation—"research and development."

Management, thinking in terms of improving operations by whatever means, should find the systems study or analysis conducted according to formalized techniques of scientific management invaluable in approaching significant procedural adjustments and a system design "molding materials, information, men, and machines into an integrated system." [3]

The case for systems study or analysis for libraries is not better stated than by Lynch of the University of Sheffield: "Systems analysis is a most valuable exercise in its own right, because it strips procedures of their mantle of age and respectability and reveals the bare bones of the scheme, and it is often possible to gain insight into and improve on a procedure without resorting to mechanization at all." [4]

The systems approach is management concentrating as responsible management continuously should, "on the analysis and design of the whole, as distinct from the analysis and design of the components or parts" [5] of systems.

Before outlining the planning of a systems study, some of the causes of the increased complexity in library management briefly alluded to earlier are more specifically explored. Some reasons for these problems, perhaps the most demanding, are suggested.

1. *The Increased Quantity and Sophistication of the Demands of Library Users.* The academic library, for example, is experiencing this impact stemming from surging enrollments; from changing teaching methods causing changing attitudes towards study; from upper-level graduate work; and from expanding research emphases. The extension of independent study programs and growing flexibility in curricular offerings and in student selection of curricula are causing more widespread and intensive use of library materials. The development of interdisciplinary study and research, requiring larger and larger collections of specialized knowledge, at the same time require application of more precise methods in selection, acquisition, processing, and control.

The effect of research activities, continuing their extensive expansion, results in two primary problems for libraries: more and more literature for the library to control and, with this, greater and greater demand for

[3] M. F. Lynch, The Library and the Computer, in *The Library and the Machine,* ed. C. D. Batty, North Midland Branch of the Library Association, Scunthorpe (Lincs.), 1966, p. 28.

[4] *Ibid.,* p. 28.

[5] Simon Ramo, *Cure for Chaos,* David McKay, New York, 1969, p. 11.

the results of research in this growing literature. Fragmentation of familiar subjects into narrower and more highly specialized categories also is causing added problems in bibliographic control and service to the researcher at the time he needs to know. Interdisciplinary or team research, to cite an example, is particularly demanding, involving as it does intermeshed specialties in the physical and the life sciences. The special librarian probably feels the urgency of the situation keenly where his company's competitive advantage may depend on the outcome of research.

This growing demand is being experienced in all types of libraries—public, special, and academic. Consequently procedures have to be upgraded and subjected to continuous study in order to make adjustments in time to meet the ever-changing and growing demands of the library's users.

2. *The Substantial Increases in Library Book Funds.* In the university and college field the aggregate of budgets for books, periodicals, and binding rose from $25,449,000 in 1956 to an estimated $111,000,000 in 1966—better than a fourfold increase for the decade. This record is almost paralleled by the public library. Special library budgets for library materials increased from $7,097,000 in 1956 to $29,152,000 in 1965; again over a fourfold increase in these years.[6] Further, substantial increase in book funds can be anticipated under continuing Federal grant programs for the development of libraries and their collections.

We need not belabor the effect of these increased funds. Unless increasingly efficient data processing systems are developed, the maintenance of effective library services may be impossible if, as reported, some $300,000,000 will go toward the acquisition of library materials by academic libraries alone by 1975, approximately three times their aggregate budgets of 1966.[7]

3. *The Increase in Interinstitutional Cooperation.* The decision on the part of a library to participate in regional, national, and international information networks brings profound changes in procedures, staff, and service requirements, and in the volume and character of data handling activities.

The implications of cooperative activities are increased funds for library materials above the regular budget; cooperation in research-materials acquisitions programs for interinstitutional service activities; and participation in area bibliographic projects and large central "processing" units.

Implications further include the library's necessary ability to conform

6 American Library Association, *Library Statistics of Colleges and Universities, 1965-66 . . .,* ALA, Chicago, 1967, p. 9, Table B; and *The Bowker Annual . . . 1968,* Bowker, New York, 1968, p. 5, Table 3A.

7 *The Bowker Annual . . . 1967,* Bowker, New York, 1967, p. 22.

to standard systems, typified by the Library of Congress *MA*chine *R*eadable *C*ataloging data (MARC II) program; to furnish speedier service to its own users as well as to the clientele of cooperating institutions; to serve a larger and more demanding user population; and to accept assignment or responsibility for specialized literature strengths as a referral center in the cooperative enterprise.

All of these factors require study with respect to the functions, procedures, and goals of every library, whether it be academic, public, or special, irrespective of the present size of the library. Every library must evaluate its current procedures, from acquisitions to reader services, if it is to fit into the modern concepts of evolving library networks.

4. *The Shortage of Professional Librarians.* The widely felt shortage of professional librarians may be traceable to a greater extent than known to the many misused man-hours in which librarians are absorbed in clerical functions. Application of integrated data processing procedures based on the recommendations of a systems study should result in the clerk or the computer taking over appropriate tasks, thus freeing the librarian for creative endeavor.

What are the principal symptoms of a library's inability to adapt to the pressures and increasing demands implicit in the foregoing reasons causing complexities in the library? These symptoms are identified by (a) increasing unit costs in processing library materials; (b) increasing backlog of requests and of materials standing unprocessed for unidentified reasons; and (c) deterioration in the services needed and expected by the library's community.

It is the responsibility of library management to evaluate continually each of the operating systems for determining whether any one or a combination of the preceding symptoms exists. In order to do this library management should be capable of rudimentary analysis and evaluation of data processing systems. Lacking this skill or ability or waiting too long may result in the breakdown of an individual system and conceivably eventually of the total library system. It simply is not sufficient that management merely recognize that problems exist.

It was recognition of this responsibility by the library management in cooperation with the administration of Rensselaer that led to a systems study and to the development of the methodology and procedures covered by this guide. In 1965 the services of a professional management consulting organization were employed to make a study including the feasibility of machine-oriented procedures. Since there were serious misgivings about the efficacy of the recommendations flowing from this study, it was decided to employ a management systems analyst as a full-time library staff member to study the suggested changes. The analyst's first action was to review the recommendations thoroughly. The complete inadequacy of the recommendations of this 1965 survey conducted *seriatim* with the consultant not having taken enough time for acquiring an understanding of the complexities of the library's problems was recognized. This situation well demonstrated the principle that such a study cannot be successfully prosecuted except by one with enough time to become fully immersed in the library in all respects.

After some six months of intensive study of the library's systems and of the demands and requirements being placed on them, the analyst submitted a new written proposal for the library's "rehabilitation." Administrative approval to implement the proposal was received in November, 1965. From the beginning the study's recommendations were oriented to a computer-based system. Factors of long-range planning as well as the availability of a computer center led to this approach.

Although pertinent to the preliminary stages of the design of a computer-based system, this guide is not intended to prepare one for complete design of such a system. It can, however, engender the ability to perform at least these functions:

1. To analyze and evaluate most basic manual methods and procedures and to design or modify them for smoother and more efficient operation.

2. To discuss intelligently with an outside systems analyst, computer specialist, and programmer the information each requires; and to understand better what each will be looking for in studying the library's systems and in designing new ones.

3. To evaluate the findings and recommendations of an analyst and to determine whether implementation of such recommendations will, in fact, satisfy the requirements specified for the system by library management.

As a "valuable exercise in its own right" the systems study assures library management that decisions made to correct the problems now and in the predictable future are based on rigorous research conducted according to the principles of scientific management.

chapter one

SYSTEMS IN A LIBRARY

A dictionary definition of a system is "an assemblage of objects united by some form of regular *interaction* or *interdependence*." This definition implies dynamic processes consisting of interrelated functions that tend to become increasingly complex as the volume of requirements placed on a system grows.

TYPES OF LIBRARY SYSTEMS

Broadly conceived the library functions in the framework of two major types of systems: the data processing and the informational. The data processing system may be defined as the organization and the methods involved to perform operations necessary to effect the form or content of information needed to satisfy the library's *management requirements* and goals.

The informational system may be defined as the organization and the methods followed in storing and retrieving information to satisfy the library's *service requirements* and goals.

Generalizing the distinction between the two major types of systems, the data processing system is concerned with the manipulation of data (clerical functions) and the informational system with the storage of information and its recovery.

SIX BASIC LIBRARY SYSTEMS

On a more specific level six basic systems can be identified within the library. All contain the characteristics of the data processing and informational systems previously defined. Some of these belong almost completely in one category, while others consist of a combination of data processing and information storage and recovery.

Considering its predominant functions of ordering and receiving library materials and paying and accounting, the functions of the *acquisitions system* depend mainly on data processing operations. Likewise the *serials control* system is heavily concerned with data processing functions from which, however, an informational system evolves that permanently stores serials information for future reference.

The *circulation control* and the *library administration and planning* systems are examples of the combination of the data processing and the informational systems. The major function of the circulation system is the control of the flow and movement of library materials. The function of the administrative system is that of organizing and controlling the operation of the library as a whole, receiving the reports and statistics of each of the other systems, and summarizing and analyzing this information to make meaningful decisions and to determine whether management goals are being served.

The *cataloging system,* with an appreciable amount of data processing, is primarily an informational system charged with classifying books and other library materials and providing the records essential for retrieving them. The *reference system* also may be regarded as an informational system, concerned as it is with retrieval and transfer of information required by the library user.

ELEMENTS OF A SYSTEM

In any type of system will be found four basic elements: the input, storage, processing, and output components. That these four components are present in any system is demonstrated by a random illustration of input, storage, processing, and output of the human "nervous system": *input,* finger touches a hot stove; *storage,* the impulse arrives at the brain where the previous experiences of the system as a whole are stored; *processing,* the brain sorts out the questions asked of it and transfers these to a pain signal; *output,* the pain signal triggers the removal of the finger from the stove.

We can consider the data processing and informational systems referred to as the two major types in a library. The data processing system contains these four elements: the *input element;* the *memory element* (storage), or unit, where information is stored until it is ready to be processed; the *processing element* where the various calculations necessary to transactions are made; and finally the *output element* where the desired results are obtained or action occurs. It will be recognized that these four elements in data processing are identical to those found in a computer system.

The informational system is almost identical to the data processing system in the generalized view. It too contains the *input* unit or element by which information is received; the *memory* unit (storage) where the information is held until desired; and the *output* unit. There is no need for the *processing* unit. The information is required with no manipulation or change.

The formation and use of the traditional library card catalog illustrate the operation of an informational system. The inputs are the cards prepared and filed by the cataloging department, and the memory and storage unit is the file in which the catalog cards are stored. In this example the output element (the catalog card) is not activated until a reader approaches the system, the card catalog, with a need for information on a specified subject, thus generating an input that is his need for information. The informational system cycle is continued: the reader locates in the catalog the desired information—output—that the book wanted is in the library and that it carries a specific location number.

SUBSYSTEMS

The model of an acquisition system (Figure 1-1) serves as an example for analyses of interrelated functions within a system. The figure shows the combination of all the components required to make this acquisitions system operational. Within this or any other system there is always a group of interrelated subsystems, each of which is designed for the performance of a particular task. This acquisitions system model consists of the subsystems *preorder search, ordering, receiving and checking,* and *accounting and reporting.* The subsystems depend one on the other for input and output to fulfill specified requirements. For example, the order subsystem could not operate alone; it requires the verified information (output) supplied by the preorder search subsystem.

The detailed illustration in Figure 1-1 of one phase of work within the receiving and checking subsystem of the acquisitions system serves to emphasize the interface between and among systems. The operation of checking in books carries with it the function of comparing the on-order slip with the book when it is received. A decision is made on whether or

FIGURE 1-1 Model of an acquisitions system showing one phase of the interdependence of systems

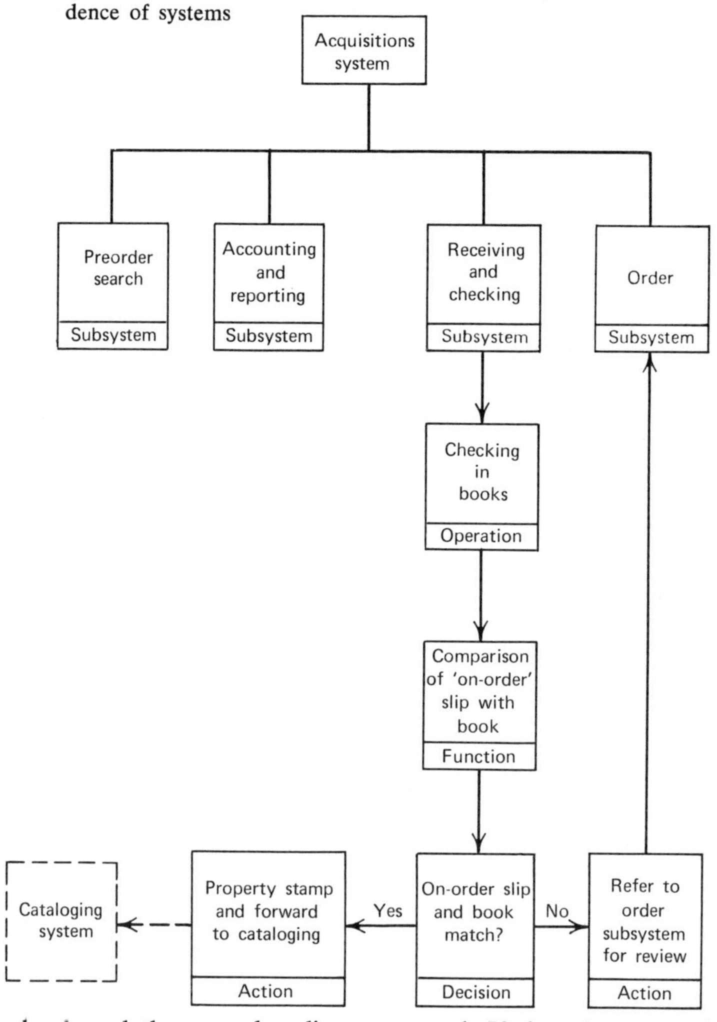

not the book and the on-order slip correspond. If they do not, action is taken so that the slip and the book are referred to the order subsystem for review and follow-up. If the two do correspond, the resulting action is the property stamping of the book and its being forwarded to the cataloging system for further processing.

OPERATIONS OF A SYSTEM

Each subsystem consists of a group of logically interrelated operations. Each operation is concerned with the performance of certain functions, or certain decisions, or both, as the bases of action. For example, the preorder search subsystem may consist of the following illustrative elements:

Operations
Searching library records
Searching trade bibliographies and other sources of title verification
(a) *Functions*
Verification of title
Provision of order information
(b) *Decisions and Actions*
Decision: Is item in collection? Yes........ No........
Action: Duplicate........ Do not duplicate........
Decision: Is item available? Yes........ No........
Action: Order........ Not order........
Decision: Information sufficient to satisfy ordering subsystem? Yes........ No........
Action: Order........ Refer to requestor for additional information........

Thus we can see that the *operations,* which are the major elements of the subsystem, consist of specified *functions, decisions,* and *actions,* also known and referred to as *elements of operations*—the basic "building blocks" of a system and its subsystems. In an effort to clarify the meaning of and relationships among these elements the following explanations are offered: (a) *operations*—processing; (b) *functions*—the substance of processing, that is, the objectives; (c) *decisions*—the questions to be answered in the course of the processing; and (d) *action*—action taken as a result of the decision.

Consistency of approach in the study of each system based on these elements reveals the interaction among operations in all systems of the library organization.

TOTAL SYSTEMS

There is no system in which the operations are isolated within a single subsystem. Characteristically, as illustrated, it will be found that there is a direct as well as an indirect effect on the operations of another system and its subsystems. It is the complexity of this interaction between and among systems that is causing libraries to investigate the concept of "total systems" before automating any operation when undertaking a systems study.

When synthesized and coordinated, the six basic library systems comprise what may be termed the "total library system" or the library functioning as an entity (Figure 1-2). It is the combination of these systems, with

FIGURE 1-2 Model of a total library system

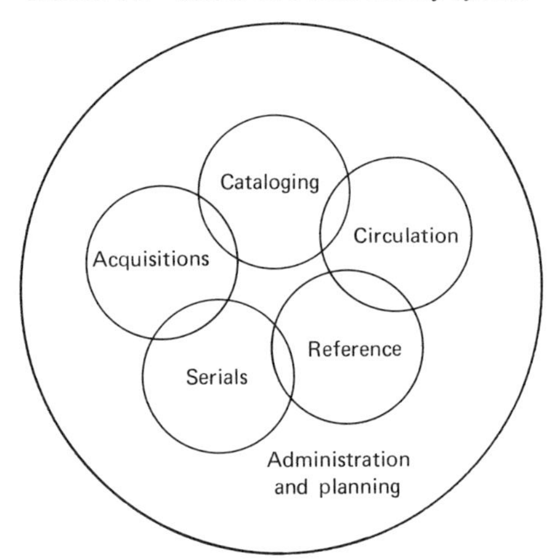

well-defined operational interrelationships and recognition of information transfer requirements among them, that assures effective and efficient operations directed toward achievement of the total system's, the library's overall established goals. Integration is necessary to facilitate exchange of significant information on a need-to-know basis among systems comprising the total system. A total system is one in which traditional departmental boundaries tend to fade and artificial separations find no place.

The total systems concept is not new in library administration but its full application beyond acknowledgement of it through a broadly functional organizational structure has been hampered and, indeed, prevented by the absence and unavailability of technological aids for a job that has always been overwhelming—the storage and controlled use of the mass of "data" native to the library. The use of modern business methods for processing and storing the varied, accumulating, and ever-changing data generated and used by the library would appear to be a requirement in the medium-to-large-sized library.

The total systems concept is more completely and successfully implemented in terms of preciseness and timeliness of needed information through machine methods than by laborious, economically unjustified, and error-prone manual methods. To be sure, reference personnel in evaluating the library's collection can hand produce subject area lists for analysis. More probably such lists will be generated by merely scanning the shelf list under main classifications, resulting in an imprecise, generalized conclusion regarding the improvement of the collection. On the other hand by using data generated by the acquisitions and cataloging departments and stored in computer memory the reference personnel can study this storage to obtain accurate and detailed information and then use their time applying the professional judgement necessary for valid collection evaluation. Examples of the high utility of the computer in applying the total systems principle can be multiplied. Circulation control based on the data stored in the computer by acquisitions and cataloging becomes more precise and timely and can be programmed for ready analysis of collection use and reader characteristics. This can be done manually but the results are likely to be reported too late and at too great a cost to be useful in improving currently chronic service deficiencies.

FEEDBACK

The principle that has led to the application of the total systems concept is that of monitoring and feedback.[1] In a total system feedback is the relaying of information from one system to another, so that a given system can adjust to satisfy the requirements of the total system. For example if the demand for books in a particular subject area is increasing, this is evidenced in the circulation system's records. In a computer based system the information in these records would be monitored. When demand exceeds predetermined criteria, management would be notified for the purpose of taking appropriate action.

Therefore planning of individual systems is not done in isolation. Such planning should always be in the context of developing a total system for giving maximum service to readers and eliminating duplication of staff effort.

THE LIBRARY DATA PROCESSING SYSTEM

The complexity of any data processing system prohibits analysis, evaluation, and design as a one-step process (see Chapter 2). The job of the

[1] For a business management definition of feedback see: S. L. Optner, *Systems Analysis for Busines Management,* Prentice-Hall, Englewood Cliffs, N.J., 1968, pp. 40-42.

study staff in the analysis phase of the data processing system is that of identifying the major activities of the system, the *subsystems* (see Chapter 3). The subsystems together satisfy one or more secondary requirements or demands placed on the system and contribute to the satisfaction of the system's primary requirements. With the subsystems identified the operations occurring within the subsystem can be isolated and the functions and decisions associated with the manipulation of each input needed to prepare the desired output can be defined. Finally the required actions resulting from the various decisions also can be defined.

FIGURE 1-3 Model of elements within a system

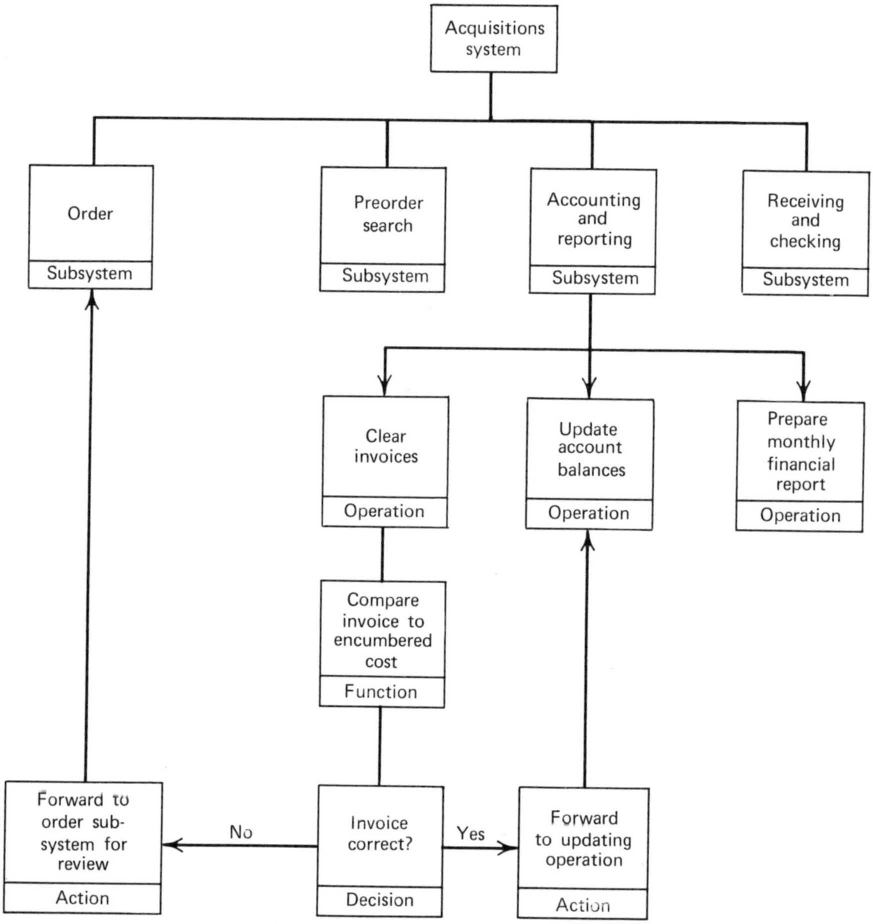

The functions and decisions are the basic elements of a data processing system. The evaluation of a system consists of determining how well the functions are organized and performed and how appropriate the decisions are in satisfying the requirements of the system. In addition the design of a new system should start with these basic "building blocks," the elements. Figure 1-3 illustrates the elements in a library data processing system, an acquisitions system. Let us analyze two operations in the accounting and reporting subsystem (Figure 1-4).

FIGURE 1-4 Accounting and reporting subsystem operations

Operation	Function	Decision	Action	
x				Clear invoices for payment
	x			Compare invoice to encumbered cost
		x		Is invoice correct or complete?
			x	Yes: clear invoice and forward to account updating operation
			x	No: forward invoice to order subsystem for review
x				Update account balances
	x			Assign proper account number to invoice
	x			Post the order, invoice and account numbers, and amount to invoice register
	x			Maintain current balance on each book fund
	x			Post invoice amount to proper book fund
	x			Calculate new balance for each book fund
		x		Has book fund reached minimum balance?
			x	Yes: notify library director for authorization of additional funds or discontinue ordering
			x	No: continue processing of subsequent invoices

The level of detail in such an analysis as this depends on the type of system being developed. If the design of a computer system is contemplated, the functions and decisions must be defined in sufficient detail to permit a programmer to program the system.

THE TOTAL SYSTEMS CONCEPT REVIEWED

The total system operates within an environment dictated by the goals of its parent organization or governing body.

"The first consideration in the total systems concept is description of its objectives. They can be simply stated as (a) to organize administrative work flow from the viewpoint of the [library] as a whole, without regard for barriers of organizational segments; and (b) to develop data processing systems whereby source data are recorded once and thereafter perpetuated in various summary forms to meet departmental operating and financial needs, without repetitive processing." [2]

The realization of the total system concept occurs when the major operating systems have been so designed and integrated that the interaction and exchange of information are facilitated in a logical and systematic manner. The planning of individual systems is not an end in itself. Such planning must always be in the context of developing a total system of reciprocal relationships among individual systems. This requires the design of integrated procedures, taking into account all of the library's systems.

The cost of punched card equipment, or computer equipment, or both, and the extent to which this equipment can profitably be put to use for any library requires planning for the future. The total systems concept can be adapted to limited use of machines for appropriate operations within a system or for all operations in one or several systems. The urgency is that a systems study be oriented toward a future goal and that the six basic library systems be tested, reviewed, and designed so that all can perform as a whole. The decision to use machines is an administrative one and depends on all the factors that determine the role of any one library. Although a computer-based total systems goal may be a good administrative objective, the systems study could be hampered by the question of the administration's economic ability. It is now generally realized that computer use rarely saves money. It is more important to recognize that proper use of the

[2] J. W. Haslett, Total Systems—A Concept of Procedural Relationships in Information Processing, in *Total Systems,* ed. Alan D. Meacham and Van B. Thompson, American Data Processing, Inc., Detroit, Mich., 1962, p. 16.

special talents of professionally trained librarians is essential to handle the proliferating volume of literature to be controlled, so that the users can be more profitably served. If machines are not the answer for a small library, the study should result in the design of operations in order that clerical processing can be done by other than the professional staff members who will then be free to render the creative and imaginative service appropriate to their professional qualifications.

chapter two

PLANNING AND CONDUCTING SYSTEMS STUDY[1]

In the Introduction the causes of the increased complexity in libraries have been given in some detail. A systems study is indispensable when a library faces these problems and admits that it is no longer serving its community effectively.

"Systems study" is equated to the phrase "let's get organized." It is prerequisite to a well-designed and successful automated system but the results of such a study need not be an automated system. Study methods and techniques can and should be used in analyzing, evaluating, and designing all levels of data processing. The librarian familiar with the concepts and techniques of systems study should be able to increase the efficiency and productivity of the library even if the only available mechanical equipment is the typewriter.

[1] The following are regarded as good introductory texts: Systems and Procedures Association, *Business Systems,* The Association, Cleveland, 1966. R. F. Neuschel, *Management by System,* McGraw-Hill, New York, 1960. J. W. Greenwood, *EDP: The Feasibility Study—Analysis and Improvement of Data Processing* (Systems Education Monograph No. 4) Systems and Procedures Association, Detroit, Mich., 1962. With the Association's permission, Chapters I and II have been used for guidance in preparing this chapter on organizing, planning, and conducting a library systems study. (Systems and Procedures Association, now Association for Systems Management, 24587 Bagley Road, Cleveland, Ohio 44138.)

SYSTEMS STUDY CONCEPT

The systems study is defined as the logical analysis of the present systems; the evaluation of the efficiency, economy, accuracy, productivity, and timeliness of existing methods and procedures measured against the established goals of the library; and the design of new methods and procedures or modification of existing methods and procedures to improve the flow of information through the systems.

The main distinction between the analysis and design phases is that analysis is a rigorously controlled inquiry into existing conditions, while design is the resulting, synthesizing process in which new ideas are generated and refined. Design is the final phase of a systems study involving creative thinking, coordination of the conclusions reached in the analysis, and deductive reasoning directed toward realization of the stated objectives and goals of management.

The concept of systems study consists of three interdependent phases:

1. Analysis, which is the accurate delineation of the requirements placed on a system; the current procedures by which the requirements are met; the outputs of the system in satisfaction of the system's requirements; and the inputs used to generate the outputs. The four items under analysis represent concurrent identification of the areas of inquiry, coupled with the charting of all operations, functions, decisions, and actions, the gathering of data produced and forms used, the listing and evaluation of available personnel and equipment, all synthesized into a report of existing conditions.

2. Evaluation, which is the detailed examination of current procedures with respect to their adequacy to implement the mission of the system.

3. Design, which is the action taken by validation of the existing system, by modification of it, or by substitution of a newly designed system to satisfy the demands being placed on the system.

DEFINITION OF TERMS

The use of terms in subsequent detailing of phases of a systems study can be a source of confusion. It is our intention here to avoid technical language as much as possible in order to serve better the understanding of the nonspecialist. The simple terms chosen are applicable to processing by any means, manual or machine, and can be expanded or modified where appropriate. It is suggested that the following terms are commonly in use and therefore can be helpful in effective interchange of ideas with management specialists. The reader will note that the arrangement of the terms attempts to indicate their logical and hierarchial interrelationships.

Goals: the objectives of the total system that establish demands placed against each system and its subsystems.
Demands: the established requirements of a system and its subsystems.
Requirements: the supply of data, information, action taken resulting from a demand.
Input: the printed form, written record, oral information, instructions needed to satisfy a requirement.
Output: the answer to the requirement of a system in the form necessary to convey or transmit information.
Subsystem: a major part, component, or activity of a system.
Operations: the major, specific units of work.
Procedures: used synonomously with operations.
Jobs: used synonomously with operations and procedures.
Elements of operations: the functions, decisions, actions comprising operations.
Functions: the processing steps in operations.
Decisions: the determination of the steps to be taken to complete a function.
Action: the course taken as a result of the decision .

LIBRARIANS AND THE SYSTEMS STUDY

Intelligent staff participation and interest are essential to the success of any study to effect operating improvements in a system. In order to consolidate the gains that can come from a systems study the personnel must be capable of maintaining surveillance of the recommended operational structure and procedures.

A systems study is the beginning for a different administrative and organizational work pattern that must be monitored in order to maintain and improve the library's ability to achieve the goals set for it. Regular analysis and evaluation of current methods and procedures should ensure that the demands for information and action being placed on the systems are being met. If they are not, the design and modification of methods and procedures become necessary because of the changing requirements of the

[2] P. Wasserman, *The Librarian and the Machine,* Gale, Detroit, Mich., 1965, pp. 87-96.

system. Thus the need for the "new breed in librarianship," as Paul Wasserman puts it in his monograph "The Librarian and the Machine." [2] This "new breed" is the library systems analyst and the programming librarian. The presence of such personnel trained in the required managerial skills should release other professional librarians from processing functions to the areas of professional services to library users and to research in user requirements and improved methods of service.

Basic education in librarianship coupled with specialization in data processing, or computer programming, or both, is suggested, if not essential, for the professionally oriented library systems analyst and programming librarian. As held by both Wasserman[3] and Minder[4] the problems to be solved are not simply tied to managerial technology but also involve understanding of the professional objectives to be served. Librarians must be able to analyze, evaluate, and design their own systems in terms of the requirements of librarianship. Otherwise the management and computer technologist, unaware or uncertain of the goals of librarianship, may adversely influence the organizational structure required for the attainment of the purposes of library service. Again, as Wasserman writes in discussing the use of technological expertise by libraries, "The responsibility of the data processing person . . . must be subordinate to a management which is broader and more comprehensive in its approach to the program of the library. Even when those who are trained or experienced in data processing begin to assume general and broader library management responsibility, it must be for the ability which those so chosen manifest in these broader problems of organization than simply those tied to its technology." [5] Thus it seems advisable to look toward the library profession itself in the development of these new administrative positions in order to maintain balance in the composition of library administrative leadership.

It would appear desirable for a library to have on its administrative staff at least one full-time officer whose responsibility is that of improving the library's systems and procedures and constantly modifying processes and procedures as demands or requirements placed on the organization change. The size of a library is not a factor in the need for systems study. In the case of the "one-man" library it would be left for the librarian himself to acquire the requisite managerial skills. On the other hand the medium- to large-sized library organizations might well think in terms of full-time staff specialists.

[3] Wasserman, *op. cit.*, p. 88.

[4] T. Minder, Library Systems Analyst—A Job Description, *College and Research Libraries*, **27**: 274-275 (July 1966).

[5] Wasserman, *op. cit.*, p. 88.

Since the systems analyst or manager of systems and procedures is a staff officer, established and accepted as an agent of the library's director, his point of view and interest must be coterminous with those of the director. The responsibilities of this position, in cooperation with library department heads, include the following functions:[6]

1. Assisting management in the review and evaluation of operations and services to meet the established goals of the library.
2. Designing and implementing in cooperation with supervisory staff new or improved operating systems for increasing effectiveness, strengthening operating or management controls, and expediting performance of routine work.
3. Developing operating manuals and reviewing, improving, and planning statistical and accounting reports for managerial control at all levels.
4. Evaluating existing forms and as necessary designing new or improved forms.
5. Conducting training programs for staff management in the regular application of systems study techniques to daily operating problems and in the capabilities and use of the computer in operations and library services.
6. Directing the design and programming of computer-based systems and representing the library's interests in shared computer facilities.
7. Keeping abreast of new developments in data processing, together with associated equipment, and their application in library operations.

Library systems analysis, evaluation, and design will be ineffectual unless done by persons who are trained or formally educated in librarianship. Although without experience, the library school graduate is prepared to learn the nuances of library service that can only be gained through experience. At least his schooling has made him aware of this contingency and he is prepared to develop professional attitudes. This cannot be safely said of the one trained solely in managerial techniques.

Library schools in the main are beginning to recognize the need for an introduction to systems study and data processing. A recent sampling of some half dozen schools revealed thinking in this direction. This is corroborated by Hayes' table of university programs.[7] There is, however, a tendency to feel that the student librarian without sufficient prior knowledge of the components of a total library system would not properly profit from instruction in these phases of managerial science. Cannot this be said, too, of the student in such a course as "Library Administration" or "Li-

[6] Minder, *op. cit.*, pp. 274-275.

[7] Robert M. Hayes, Data Processing in the Library School Curriculum, *ALA Bulletin*, **61**: 622-668 (June 1967).

brary Organization," where the application of principles can only be inferred and likely misinterpreted in the case of the many students who have not even elementary exposure to the operation of a library?

Paul Wasserman suggests that the library school should offer programs in data processing to the practicing librarian at an intermediate level, "not necessarily tied to any formal degree level resulting in some type of certificate midway between the master's and doctorate.[8] Conversely it is suggested that programs in the specific functions of the control systems of a library be offered to management degree holders. This can be done and is being done by some libraries through in-service study programs. In institutions where both a library degree and a management degree are offered, an interdisciplinary program could be profitable in making available to libraries management personnel knowledgeable in library operations.

Modern library management, aware of the need for systems study, has no recourse but to learn the techniques and tools of systems study and the skills to apply them—the basis of good management (Moore;[9] Schultheis[10]). If a library is to examine itself with the techniques of management science, the responsible personnel should be trained to do so.

Because the systems study represents a demanding total library effort that may result in major operating changes, the entire library staff, under the strong leadership of the administration of the library, should be fully involved in planning and conducting the study. Here the guidance, direction, and personal participation of the director of the library are critical. He is responsible for the proper execution of the study.

STUDY STAFF

Having received governing administrative authorization to proceed, the first responsibility of the director of the library is to select and appoint the staff to make the study. There are two principles to be kept in mind here: (a) a good systems study cannot be done on a part-time basis; and (b) the person selected to direct the study preferably should possess a combination of education in librarianship and training in the methods of systems analysis as taught in modern management courses. Since satisfac-

[8] Wasserman, *op. cit.*, p. 138.

[9] E. Moore, Systems Analysis: An Overview, *Special Libraries,* **58**:87-90 (February 1967).

[10] L. Schultheis, Systems Analysis and Planning, in *Data Processing in Public and University Libraries,* ed. John Harvey, Spartan, Washington, D.C., 1966, pp. 96-102.

tion of the latter prerequisite probably is not possible in most instances, one of two courses is open: either to release a library staff member from other duties to prepare himself for the conduct of a study, or, to bring in a skilled systems analyst unfamiliar with library organization.

The outside analyst skilled in modern management techniques should become fully familiar with the library's problems and responsibilities and develop a rapport and identification particularly with the supervisory staff. It is probable that the analyst from outside the library organization will require at least six months of intensive study and orientation to become sufficiently familiar with the library's problems and responsibilities and to develop rapport with the staff. He should not approach the study with preconceived notions of the local problems to be solved, possibly arising from his work with another library or with what he feels is an analogous organization. Although the conditions he sees in a cursory inspection may appear to be the same as he has encountered before, the causes of the conditions to be studied frequently are completely different. The analyst's responsibility is to determine *what* problems exist, *how* and *where* they originated and *how* the conditions at hand can be corrected.

The composition of the study staff should be as follows:

1. A library officer possessing responsibility and authority within the library's organization, to take supervision of the study.

2. At least one member of the study staff fully trained and experienced in the application of management-analysis techniques, preferably a librarian.

3. At least one member of the staff skilled in electronic data processing (EDP) methods, particularly if an automated system is contemplated; and again this member preferably should be a librarian.

4. Clerical assistance sufficient to support the work of the study staff.

It is conceivable that the requirements in Items 1-3 could be met by one person. However, it is more likely that two persons will be needed to meet the indicated requirements in the case of the medium- or large-sized libraries. A major factor here, of course, is that the library must continue to operate during the study and the dual responsibility could impede the progress of the study if only one person undertakes the role.

DEFINING THE STUDY PROBLEMS

The study staff is to develop with the library's management a detailed procedural plan and time schedule for the study, the first step of which is the general definition of the problems to be studied and the identification and description of the specific problems involved. It is the responsibility of the library's management to supply a clear and concise statement of these problems in the form of a *written report* such as would have been prepared for gaining governing administrative support and financing for the systems study originally (see Appendix to this chapter). Problem definition must be in sufficient detail to serve as a guide to the study staff members and to inform the other members of the library's organization who are concerned with the activities named as areas to be covered by the study.

LONG-TERM GOALS

The next step, critical in planning and conducting the study, is the definition of the library's overall goals, of which the systems study staff must be fully aware in relation to the problems to be studied. Goals are those factors that the management of the library determine to be important for accomplishment by the library. The statement of goals determines the major requirements resulting from demands that should be satisfied by the library through its data processing and informational systems. Goals are basic in evaluating the current system and in designing a new system. If the goals are not precisely and correctly defined and understood in detail, the results can only be an inaccurate evaluation of current operations and the design of a faulty system.

Therefore the study staff should verify the validity of the stated goals through discussion with library users, department heads in the library, department heads of the parent organization of the library, and with the key administrative officers of the library's governing organization.

Long-term planning is implicit in defining the library's goals. Because most libraries are in a period of dynamic and persistent growth, systems to satisfy long-term goals should be designed with the capability of handling increased demands. Thus computer-based systems are suggested. Such systems have the capability of less costly and more efficient growth as demands on the library grow. Today's commonly applied manual systems often become more costly and less efficient under increasing demand and do not possess the "stand-by" qualities of the computer system in adjusting to growing requirements.

The following are some of the major goals of many libraries today: (a) to improve service to users; (b) to provide more prompt access and greater use of the library's resources; and (c) to participate in library network programs of bibliographic data distribution and information transfer.

SCOPE

After the goals have been defined, the scope of the study must be established. Here the particular data processing systems within the library that are to be studied are identified as well as the organizational units in which the operations are performed and the activities involved in order to prevent wandering into other systems with which management is not concerned at this juncture; and within the system priorities are established for the components thought to need early attention by the library's administration. It is important in specifying scope and priorities that the systems study staff not be too rigidly restricted but rather allowed a degree of flexibility permitting recognition of other areas that might be affected by the particular system directly assigned for study.

LIMITS AND RESTRICTIONS

Within the scope management also should define any limits or restrictions to be placed on the development of a system. It is important, obviously, for the study staff to know these parameters before the systems design phase. Illustrative of limits and restrictions are (a) the type of system wanted: computer based or not computer based; (b) number and proportional distribution of personnel to be in the system; and (c) tolerable unit or total costs of operations of the system.

METHODS, TECHNIQUES

In further preparation the study staff members should decide on the methods and techniques to be used in the study for obtaining and recording the necessary information. These should be determined in order to assure a logical, systematic study and to permit comparison of findings within the systems. For example, if statistical work sampling is to be used, the confidence level to be accepted must be agreed on in advance with applicable sampling techniques being uniformly employed; survey forms must be designed for each of the surveys to be conducted; type and format of reports to be presented must be determined—that is, graphic reports such as organization charts, procedure flow charts, graphs, tables, and so forth; and written reports or other narrative materials must be agreed on. Involved also in the methods used are designation of the persons to be interviewed, the records to be obtained and analyzed, the equipment available and its use, and a detailed outline of the specific types of information to be sought. Adoption of uniform methods and procedures assures that the results reported by each study staff member are consistent in content and format and allow uniform comparison and evaluation by the staff in consultation.

WORK AND TIME SCHEDULE

As terminal planning steps the organization of the work of the study staff should be set down and a time schedule prepared for completion of its assignment measured by man-days or man-months; the types of skills needed for each assignment have to be determined—that is, managerial analysis, clerical, programming, and so forth; the responsibilities of each person in prosecuting and completing his study assignment should be explicitly defined based on a list of specific identifiable study stages, each with a target date for completion; target dates for interim reporting to the library's management and to the study group as a whole should be set as well as the target date for submission of the final study report. The time schedule is not only important to the orderly and expeditious prosecution of the study but also to administrative knowledge and acceptance of how long current library operations will be slowed or otherwise adversely affected by the study's demands on the operating time.[11]

ANNOUNCEMENT OF STUDY PLAN

As the plan for the systems study is being formulated the director of the library should have introduced the idea to the library staff, indicating the reasons for the study and its objectives. Such announcement would include (a) an indication of the reasons for the study; (b) a description of the principal goals of the study; (c) a description of the benefits expected; and (d) solicitation of the full cooperation of the members of the library staff, assuring them of their major roles and engendering their full support and interest in order that the study be successful.

When the planning is completed, the director should review the finished plan with the study staff and with other members of the library's administration. When he has approved the plan, the director of the library should make a formal announcement of the undertaking of the systems study to

[11] G. W. Covill, Librarian and Systems Analyst = Teamwork?, *Special Libraries*, **58**: 99-101 (February 1967).

the community served by the library, indicating the reasons for the study and its objectives and anticipated benefits as well as explaining what problems the user of the library may temporarily encounter during the period of the study.

The library staff should be reassured at this time of the administration's awareness and sympathetic understanding of the disruption of each staff member's assigned duties and, further, be assured of the administration's firm support of the study. It should be demonstrated to the staff that without the cooperation and participation of each member the study cannot lead to results beneficial to the staff and to the library.

STAFF TRAINING PROGRAM

Following the announcement of the starting date of the study, it would be well to conduct a short staff training program detailing the techniques used in a systems study. Discussion of these techniques should bring a better understanding of the study, generate the very necessary staff interest in it, and furnish knowledge of what information the analyst will be seeking in his contacts with individual library staff members. The training program also might well include description of the potential use and contribution of computers in library data processing operations and library services. At the same time the library staff should be provided with selected references to the current literature on library automation.

It cannot be overstressed that staff understanding, perceptiveness, and support are prerequisite to a systems study resulting in practicable systems design. Support in this phase of organizational improvement efforts will be carried over into the successful implementation of the study's recommendations. As in any activity staff cooperation, willingness, interest, desire, and pride in work generally assures attainment of operating objectives.

APPENDIX

To: Responsible Administrative Officer
From: Librarian
Subject: Current Condition of the Library and Request for Detailed Study of the Library System.

This memorandum is a report on the condition of the library and the resulting problems affecting the efficiency of operations and realization of the library's ultimate goal of the best possible service to its users.

The purpose of this report is to request approval for a full analysis of the library's problems to determine the cost of correcting them and the cost of maintaining and improving on the resulting gains.

The analysis would be undertaken with the full cooperation of the present library staff which at the same time must maintain operations and services at the present level to the extent possible while supplying the data needed in the study. For this reason I believe a detailed survey and analysis requires the assistance of a systems analyst whose sole function would be to study the problems and recommend actions for the improvement of the library's operations and service. Therefore I also request approval for the appointment to the library staff of a qualified analyst to conduct the study under my direction.

The problems to be investigated and solved are suggested by the conditions described below. From routinely kept operating statistics and observation some of the conditions adversely affecting the library's operations and causing a continuous decline in its ability to serve the purposes for which it exists are listed.

1. Missing from the active book collection are 5531 volumes which should be replaced.
2. The lesser used collection of 4500 books has had to be stored in cartons making the books inaccessible to users.
3. Approximately 1000 volumes of books are in serious need of binding with another 3000 requiring simple mending and relettering.
4. In the matter of the library's catalog records:
   (a) The present shelf list is not a complete listing of the library's resources nor does it provide in many instances the accurate descriptive information necessary for the identification and location of the individual items in the collections.
   (b) Over 10,000 of the shelf list records are not unit cards but merely brief and incomplete records.
   (c) The shelf list card drawers are badly overcrowded.
   (d) The public catalog is not a union catalog of all library materials in the central and outlying collections.
   (e) There are many inconsistencies of filing in the public catalog with neither the ALA or Library of Congress filing rules having been applied consistently.
   (f) The public catalog record of multivolume holdings has not been kept up in many instances.
   (g) Rules of entry for making personal and corporate name references in the public catalog have not been followed in the majority of cases.
   (h) The public catalog files are overcrowded for efficient filing and convenient consultation.

5. In the matter of the library's serials or journal collection:
   (a) More than 450 journal titles are incomplete—missing issues and volumes—largely traceable to lack of binding.
   (b) Less than 30 percent of the journal collection of 42,000 volumes is bound.
   (c) Many unbound journals are beyond repair or binding and should be replaced or discarded.
   (d) The majority of serials are only classified, not cataloged, resulting in incomplete information in the shelf list and main catalog.
   (e) Over 300 titles of monographs in series are classified in the serials collection, representing about 4000 separate monographic titles unavailable to the user by author, title, editor, subject content, and so forth.
6. In the matter of the library's ordering or acquisitions operations the backlog of user requests has increased to some 4000 items with each month's report showing a 150 to 200 growth of this backlog. Under these circumstances unit processing costs continue to rise.
7. In the matter of the library's cataloging operations:
   (a) Cataloging arrearage is now at 3500 volumes and increasing monthly.
   (b) Filing of the catalog records cannot be kept up to date.
   (c) Needed titles remain in the cataloging department well over three months on the average awaiting receipt of Library of Congress catalog card sets.
   (d) Cataloging processing is seriously slowed by use of the Dewey classification system, requiring reclassification of whole sections of older books as new editions of this classification schedule appear.
   (e) Unit costs of cataloging seem inordinately high and increasing with direct charges per unit cataloged in the range of $5.00 to $6.00.
8. In the matter of the library's reader services:
   (a) User complaints are growing, stemming largely from the delay in book ordering and cataloging.
   (b) The library is impaired in its capability to participate in cooperative network services due mainly to the inadequacy and lack of standardization of its bibliographic records.
   (c) The reader services staff is handicapped in providing services to our users as a result of the incomplete and inaccurate catalog and other bibliographic records.
   (d) Circulation records are in many instances not useful in identifying the location of materials out of the central collection due to the large number of missing books and those placed in inactive, inaccessible storage.
   (e) Many user complaints are based on not finding materials reportedly in the collection. Many users report finding only one out of every three books they are seeking.

(f) Mutilation of journals and books is evident and the necessarily slow replacement of missing pages and issues is another source of user complaints.

9. In the matter of the branch libraries:
   (a) Thirty-eight percent of the central library book budget was spent in duplicating materials in the branch libraries during the past fiscal year.
   (b) Various shelving, classification, and cataloging schemes are employed by the branch libraries. Some do not maintain card catalogs or indexes of any sort.
   (c) Circulation policies in the branch libraries are capricious. Certain ones allow books to circulate for various lengths of time with little if any follow-up procedure, while others allow no book check-out at all.

The existing condition of the library is the result of attempts over a period of years to satisfy the needs of our community with a limited operating budget and the consequent stop-gap measures. Many libraries today find themselves in this position and it appears to me that we would be shortsighted if we did not study our current problems and plan for the demands of the future by undertaking a total systems study.

The study should lead to establishing the interrelationships of operations, and procedures could be designed that would solve the aforementioned problems as a whole rather than isolated solutions that of themselves are likely to create problems in other areas. A study may lead to an indication of the need for automation of the extensive data processing operations of a library. With our computer center it would seem logical to think of systems design in this context, whether for immediate implementation or in the future.

I believe the requested study would prepare the library to serve the immediate and long-range goals and objectives of our institution. I am aware that this proposal requires an extraordinary, appreciable expense but recommend this course as imperative to the development of the "good" library you envision.

Your authorization of a systems study conducted by an analyst skilled in scientific management techniques is respectfully urged.

Should you require clarification, I am prepared to discuss in greater detail the generally stated condition of the library and the implications as I see it. Thank you.

chapter three

ANALYSIS PHASE OF THE SYSTEMS STUDY—DETERMINATION AND SURVEY OF REQUIREMENTS

GENERAL OVERVIEW

In the analysis of systems, the initial phase of any study, the "analyst" or "analysts" on the study staff should acquire a complete understanding of the system to be surveyed and become thoroughly identified and conversant with all of its components. It is equally important that the analyst determine the interactions, the interrelations, existing between the system under investigation and all other systems that place demands on it. The analysis phase, as briefly outlined in Chapter 2, consists of four distinct but interacting surveys.

1. *The Survey of Requirements.* The requirements of the system are the results of the *demands* (for information, reports, and action) placed on the system from all sources. The analyst should determine the demands in the light of stated goals and also where they originate and how they are satisfied by the system. Are there requirements that are not satisfied and if so, why not?

2. *The Survey of Current Operating Conditions.* The analyst should obtain a working knowledge of operations required in the system, the sequence in which they are performed, the functions, decisions, and actions required for the performance of each operation, and of what the inputs and outputs of each operation are. He should make a survey of equipment to establish the utilization and capabilities of available equipment and make a detailed survey, including work measurement, of the individual jobs in the system to learn whether staff job descriptions match the functions required.

3. *The Survey of Outputs.* The outputs of a system consist of the reports, records, and actions that are prepared or performed in the system to satisfy the demands being placed against it. The analyst should determine why, how, when, and by whom the output is prepared, what information it contains, and what functions and actions are required for preparing this output.

4. *The Survey of Inputs.* An input to the system is the information that must be used in order to generate the necessary outputs needed to satisfy the system's requirements. The analyst should determine what inputs are received; how many, how often, and where they originate; what information they contain; and what functions, decisions, and actions are required in order to convert this information into the desired outputs.

REQUIREMENTS VERSUS DEMANDS

Requirements to be met by a system are based on requests or "demands" for specified information and action originating with management, the library user, and other sources both within the outside of the library. The questions needing to be asked are: What is required of the system? What must it do to satisfy the demands and needs of the library user, of management, and of the library's other systems? Exogenous demands must be identified and evaluated for their necessity and impact on the system and for their possible elimination if they are unnecessary and hamper fulfillment of the system's operational objectives.

Until requirements are known in detail, it is impossible to proceed with the analysis phase of the systems study. Understanding of present procedures, considered in Chapter 4, cannot occur without knowledge of the precise requirements a system is supposed to meet. Although the outputs of a system are derived from requirements, the two should not be confused. The outputs of a system are considered in Chapter 5. Outputs (information and action) flow from the demands with which the system is concerned.

We shall consider a few examples illustrating what requirements are in the operation of a system. Let us say that a fundamental demand of the serials system of the library is to provide the user with an accurate and current catalog or listing of the library's serials holdings. This demand obviously places a requirement on the serials system. We shall assume further that the demand is placed on the serials system to provide, within a reasonable time, special listings or special reports concerning the serials holdings felt to be necessary by the library's users and by the library staff itself. The complexity of this type of demand indicates that any serials system capable of handling it must be flexible enough to provide an analysis or report concerning the serials collection without prior knowledge of what specifically will be asked. This means that the serials system requires records that can be manipulated in various ways to satisfy multiple demands. Internal requirements of the serials system include the prompt claim of issues on a definite schedule, precise control of renewals of subscriptions when due, and improved accuracy and speed in the check-in of issues as received.

Another example of a demand is the preparation by the library's management of annual statistical reports for national data gathering agencies. This demand commits the director of the library to preparing the report and causes him to place demands on each system within the library to provide the data for this report. Thus this demand placed on the director causes the placement of multiple demands on systems throughout the library and the records required in each system must provide the information needed for the director's report.

Although the terms "demands" and "requirements" sometimes are used interchangeably, there is a difference in the meaning of these terms until the action taken translates a demand into a requirement. This fact may have become evident in the foregoing example of the demand for an annual statistical report placed against the library's management.

Whether demands originate from within the library or from outside, the reactive process explained above occurs. This complex of demands and requirements can be the source of a system's breakdown if it is not thoroughly controlled and analyzed with respect to the need, duplication, or overlapping of each element of the set of demands leading to requirements placed on the systems. Establishment of the need and correlation of demands and requirements is particularly important at the level of so-called "routine" operations. Here the library clerk with limited responsibility, an incomplete understanding of the process to which his work contributes, and no defined limitation of his own contribution may impose demands, or requirements, or both on a fellow worker resulting in a gradual deterioration of a process as it was originally designed.

The origin of the demand resulting in a requirement against a system has bearing on any judgement of whether the demand is unnecessary or must be accepted as a bonafide requirement of the system. Again alluding to the annual statistical report "demanded" by a national agency, this represents a demand from outside of the library over which the library has no direct control. The administration has two courses open. Either it can comply with all of the items of statistical information requested, placing additional requirements on the various systems of the library, or it can decide to comply partially using whatever statistical information is available. If this demand for a statistical report were an internal one and, for example, had originated with the library's administration itself, the management of the other systems of the library would be able to question elements of the demand with respect to the library's needs to be served and with respect to the applicability of the same information being made available in a different form from that requested by the library's administration.

Some demands placed on the library from outside sources may be questioned as to their being legitimate requirements acceptable by a system. This is illustrated by the instance of a comptroller's office of a parent organization demanding the continuance of manual procedures in the clearance of book invoices by the acquisitions department whose operations are to be automated. It may be pointed out that the computer system could furnish and summarize payment and accounting information in almost any manner required by the comptroller. The comptroller's demand may be shown to be unnecessary and uneconomical in the operations of both the library's acquisitions system and the comptroller's system. The comptroller may take the position that the library would create an exception in his commonly applied procedures that he is not readily inclined to change. Whatever the resolution with the comptroller of this outside demand, it serves as an illustration of why it is necessary to examine the validity of each demand.

Some demands from outside the library are patently arbitrary and create requirements deleterious to a system's operation. In a university library, for example, a department chairman may demand that his record of the books he has requested for purchase be kept up to date with the acquisitions department's record of action and ordering—that is, when each book is ordered, reports of delay, date received, and date cataloged. This requirement obviously would hamper the efficiency of the acquisitions system and create added costs if indeed the total demand could be met. This demand represents an unreasonable extension of an acceptable requirement; that is, notification to the requestor when the book is available for his use. Thus demands coming from any and all sources must be critically reviewed and analyzed in relation to the impact of the resulting requirements on the efficiency and mission of systems.

In analyzing or designing any system requirements cannot be established by asking what is wanted and blindly accepting stated requirements. Intrinsic requirements in the operation of a system frequently are not recognized or identifiable by the personnel responsible for fulfilling the purposes of a system. Explanation should be made of the *concept of requirements* if workers are to understand the meaning of their jobs and be able to identify for themselves the subsidiary requirements they fulfill in satisfying the primary demands of a system. A member of the acquisitions staff, if asked, might reply that the requirement of the acquisitions system is to order books, not realizing that many other requirements must be fulfilled to meet the basic demand of the ordering of books. Also a number of requirements are entailed if the serials system is to meet the primary demand of maintaining accurate and up-to-date records. Thus not only the primary demand of a system needs to be known but also all requirements contributing to the execution of the primary demand.

DETERMINATION OF REQUIREMENTS

The first step in the analysis of a system is that of determining what requirements are being placed on the system. The study staff must translate the stated goals of the library into the demands that the goals place on each system and, in addition, must identify all requirements and their sources as listed below.

1. From outside the local organization there are ALA rules for filing; rules for main entry; reports required by governmental agencies and professional groups.
2. From outside the library locally accounting information is required by central purchasing; user requests are required for information and services.
3. From within the library there are systems depending on other systems for information; the director of libraries requiring certain reports and statistics. Illustrative of the interdependence of systems the cataloging department may wish to receive from the acquisitions system certain information contributing to the cataloging of a book; in addition it may expect that the acquisitions department will have ordered the Library of Congress cards at the time the book was ordered; or that the acquisitions department

will supply the proof card or copy of the cataloging information from the *National Union Catalog* if available.

4. From within a system information is required by one subsystem from another; information required within a subsystem. For example, the ordering subsystem within the acquisitions department may place the requirement on the preorder search subsystem that it furnish the correct information about the availability of the publication, the publisher, the date of publication, edition, cost, author, and title.

The need for identifying sources arises in connection with the opportunities the study staff may or may not have in suggesting helpful modification, change, or elimination of existing requirements. Typical sources of requirements are represented by the following, whom the study staff should interview regarding the requirements each one places on the system: (a) the director of libraries; (b) the users of the library; (c) the heads of departments within the parent organization affecting library operations, such as purchasing, accounting, and so on; (d) the head of each major operating system within the library; (e) the head of the system being surveyed; and (f) the personnel within the system.

In analyzing the requirements of the system it is necessary to sift out those invalid requirements arising from artificial organizational separations as well as those of a traditional character serving vestigial needs. The magnitude of the need served by the requirement should be observed. If, for example, a record is maintained to answer a need that may or may not arise, it will be necessary to prove that this record is necessary. In many instances requirements have been perpetuated in a system and with the growth of the system those that are obsolete have not been eliminated. When the "why" of a requirement is explored, it may be found to be spurious. A simple relocation of records can lead to the elimination of a so-called "requirement." For example, an "official catalog," a main entry catalog of each title in the library and on order, was deemed essential in a technical processing area because of the distance of the public card catalog from the cataloging and acquisitions personnel. This requirement to maintain an official catalog, entailing a considerable drain on staff time, could be eliminated by relocating the public catalog closer to the acquisitions and cataloging departments and by filing open-order records directly into the public catalog. Thus a primary phase of the preorder searching operations of the acquisitions department could be satisfied in one place—the public card catalog.

In analyzing requirements the analyst plays a rigorously objective role finding out how the system operates and inevitably in the process evaluating the validity of the requirements. If, for example, the library's management wants a report every month of every account maintained by the library, the question "why" arises: Is this report needed every month? Would a report simply on those accounts that are running low at a particular time be more pertinent? Just how much information does management need? How is this information going to be used? In the matter of reports it will be found many times that various reports will present the same information and that several records within the library will duplicate this information. The obvious question is can one or more of these records/files be eliminated or combined? This is typical of what must be investigated in order to determine the practicability of retention of existing records and files.

This point is belabored in order to emphasize that *no* requirement, formal or informal, is too small to investigate and evaluate. In addition no requirement can be retained for any reason except that it is necessary to meet the valid demands made on the system.

The analyst must probe and question until he knows why the information flowing from a given requirement is needed, how it is used and where, whether used elsewhere or filed, and if so, why, until he knows every use and disposition of the information being generated by each requirement. In order to do this he must learn the content of specific jobs in depth and the purpose each is intended to serve. It is not a question of his being able to do the job being analyzed but rather to know what the job is about; to know the requirements the work is intended to meet; and to know the processes through which personnel attempt to satisfy the requirements. The extent of the analyst's responsibility is more precisely indicated by the questions that must be answered in the form, Worksheet for the Survey of Requirements, Figure 3-1.

Each system looked at in a cursory way may reveal apparent requirements but such requirements cannot actually be accepted unless it is found what the demands are that cause these requirements. For example, the primary demand placed on the acquisitions system is the ordering of books. This, of course, could be done by simply accepting a request and ordering the item desired. However, all requirements of the acquisitions system would not be fulfilled by this action that satisfies the primary demand. Such action would not satisfy the requirement of "no duplication"; it would not satisfy the requirement that if a book is on order another copy should not be ordered; and it would not satisfy the requirement that before ordering a book verified bibliographic information should be supplied. Without identifying within a given system all requirements both modifying and enforcing established primary demands, the system cannot be properly analyzed or designed.

Requirements are determined on the basis of actual need rather than on desire without any demonstrable reason. Otherwise an administrator who states his requirement as being the need for information about *all aspects* of an operation rather than for the *critical elements* of it will only find that his decisions affecting the maintenance of the effectiveness and efficiency of that operation are more difficult to make. Such a requirement ignores the principle of "management by exception," which is knowing what has not occurred as planned in an operation rather than all that has occurred. This principle will be reviewed in Chapter 7.

Knowledge of all requirements in specific detail is vital to determining the work force capable of satisfying these requirements. Awareness of the requirements will uncover the extent of duplication of work, multiplication of the same reports from various sources, and the actual need for the information being supplied. It is not infrequently the case that staff members receive reports serving no useful purpose to them but merely perpetuating a traditional referring of information based on defunct requirements. Such unnecessary requirements should be ferreted out and eliminated.

SURVEY OF REQUIREMENTS

In the beginning of this chapter were listed the people to be interviewed by the analyst, beginning with the director of the library, to determine the specific demands being placed on systems by the officers, workers, and users of the library and by those officers of the library's governing organization whose demands affect the library's operations.

The results of the interviews should be written if the analyst is to analyze systematically and to synthesize the information gathered. Availability of this information in a correlated condition is prerequisite and critical in the determinaton and survey of the "outputs" of systems, a step in the analysis phase of a systems study, subsequently treated in Chapter 5.

As a guide in interviewing and in maintaining a record of the information given by each interviewee, the form Worksheet for the Survey of Requirements is used (Figure 3-1). Before considering the use of this worksheet in illustrative detail, it is used to record the answers to questions such as the following:

1. What bibliographic, statistical and account records or reports are needed?
2. What system(s) generates the requested records or reports?
3. What information must these reports and records contain?
4. Why and how is this information used?

FIGURE 3-1

WORKSHEET FOR THE SURVEY OF REQUIREMENTS (PART 1)			
1. Prepare a copy of this worksheet for each of your requirements. 2. Attach a completed sample of each record or report.			
System: Acquisitions	Accounting and Reporting Subsystem	Analyst Name:	Date:
Name of Person Interviewed: Mrs. Jones		Position: Acquisitions assistant	
Name of Supervisor: Mrs. Brown		Position: Acquisitions librarian	
Describe the requirement: To maintain accurate and current balances on each fund and school account and expenditures and encumbrances by department and also to provide a monthly report on each of these accounts.			
Describe decisions you make in satisfying this requirement: What is the account number to which the book is charged? What is the variance between the list and delivered price of each book? Has the account balance fallen below the minimum that was established?			
Describe what action you may take as a result of your decisions: In the case of the major decision regarding the status of the account balance: notify the acquisitions head for authorization to transfer additional funds into a depleted account or discontinue ordering; or if the fund is not exhausted, continue charging against it.			
Describe the functions you perform in satisfying this requirement: By department, post the estimated price for each book to proper account. Post the actual cost, from invoice, for each book to account. Calculate the variance between estimated and actual cost. By school, post total expenditures and variance to school account and calculate balance. Prepare monthly financial report.			

5. Is the information received in usable or final form; if not, what functions must be performed to adapt it for use?

6. Is the proper information for making necessary decisions furnished; and what are these decisions and the basis of each?

7. What actions are normally taken as the result of these decisions?

The information obtained in the survey of requirements serves the purpose of supplying the basis for the analyst to arrive at the following determinations:

1. The information unconditionally required by each staff member at each level of the organization in order for him to meet his responsibilities; the origin of the stated requirements and their pertinence to the operation for which he is responsible;

2. The reports unconditionally needed by management, including their frequency and content;

3. The statistics that are unquestionably required for analysis of the workloads and efficiency of a system;

4. The information generated at each level of operation actually required by management for the making of valid decisions and the taking of actions actually contributing to the objectives of the system's functions.

Analysis and survey procedures can be somewhat confusing and overwhelming unless undertaken *seriatim,* proceeding gradually and systematically from the greater to the lesser factors being determined, analyzed, and surveyed. This technique should be followed in the analysis of requirements as well as in the analysis and understanding of current procedures and in the determination of outputs and of inputs of the system under study.

The first step is to analyze the system broadly in order to gain a general idea of the demands placed on it and the operations that the system must undertake if it is to satisfy its objectives. Having gained this general understanding and come to deductive conclusions, the analyst probes to a more specific level. This process is continued until he possesses sufficient detailed knowledge to allow him to evaluate the effectiveness and efficiency of the system and suggest an improved design that will best satisfy the overall requirements and goals of the library. Again, using an acquisitions system as an example, its major functions may be broadly typified by the designated subsystems: preorder search, order, accounting and reporting, and receiving and checking, all being subsystems in the model acquisitions system illustrated in Figure 1-1. With a knowledge of the system's primary demand and major functions, subsidiary or supporting requirements can now be sought and deduced. This step-building process continues until all of the operations, functions, decisions, and actions of the system are determined and synthesized in preparation for the factual systems' evaluation and design.

Referring to the Worksheet for the Survey of Requirements (Figure 3-1), the descriptions of requirements and of decisions, actions, and functions performed in satisfying each requirement are approached concurrently because of their close interdependence and are treated in this manner. The acquisitions system is taken as an example in an effort to clarify what information is being sought in the four sections of the form. Although an acquisitions system has many requirements placed on it, only one placed against the accounting and reporting subsystem (Figure 1-3) is used in illustration.

The requirement placed on the accounting and reporting subsystem is that of maintaining accurate and current balances for each book fund including a record of actual expenditures from each fund, estimated outstanding encumbrances, and the supply of monthly reports of the status of each fund. There are many minor decisions entering into this requirement: What is the fund to which each book will be charged? Is the list price of the book correct? What is the estimated variance between the list and delivered price of each book? The major decision is has a given account balance fallen to or below the minimum established for administrative action? The action triggered by the affirmative may be the transfer of funds to bolster the account or the suspension of further purchasing.

The functions performed in satisfying the requirement of accurate and timely balances include the posting of estimated prices of books at the time of order, posting of the actual costs on receipt of invoices, calculating the variances between estimated and actual charges, and the posting of total expenditures and estimated outstanding charges against each fund. The preparation of the monthly financial report for each book account is based on the execution of these functions.

Referring to Part 2 of the Worksheet for the Survey of Requirements (Figure 3-2), the description of each record or report said to be needed for satisfying given requirements should be described and justified with respect to its actual need and application. Each person preparing and maintaining a record or a report should describe each record or report and the role each plays in satisfying specified requirements. As illustrated in Figure 3-2 two records and one report are maintained and prepared by the accounting and reporting subsystem: departmental control records, school or account records, and the monthly financial report.

In determining requirements it is necessary to gain an understanding of a given system before moving to another. The analyst, however, will

FIGURE 3-2

WORKSHEET FOR THE SURVEY OF REQUIREMENTS (PART 2)
Describe below each bibliographical, statistical, and/or accounting record or report that you prepare or maintain to satisfy this requirement. (Use additional forms if necessary.) *Please attach a completed sample* of each record/report (on sample underline in *black* information you use; underline in *red* information you add; underline in *blue* any unnecessary information).
Departmental Control Record: An accounting control is maintained for each academic department as well as certain library accounts, such as reference, gifts, and so forth. Each book purchased is posted to its respective record. The record contains the estimated price, actual cost, variance, and total expenditures for each book. Periodically the information from this record is summarized and the totals posted to the proper school or major account in the school or account control record.
School or Account Control Record: A summary accounting record is maintained for each school or account. The purchases for each department are summarized and posted to the appropriate control record. The record contains the estimated cost, variance, and balance. This record is the basis for preparing the monthly financial report.
Financial Report: This report is prepared and distributed monthly. It contains the distribution of library funds as well as the breakdown of expenditures this month and year, to date, by department and school. It also presents the balance for each school and major account.
For each input received by you which in any way affects your fulfillment of this requirement, please complete a *Worksheet for the Survey of Inputs.*

inevitably gain knowledge of relationships between the system under study and other systems in the library. In studying the acquisitions system, for example, it may become evident that demands are placed on the acquisitions system by the reference system as well as the cataloging system. The detailing of such demands from other systems must be made in order to reach an understanding of the requirements placed on the acquisitions system. Thus in studying a system the analyst must determine factors outside of that system affecting or causing the requirements to be met by the system under study.

At whatever level in the organization requirements are being investigated much the same questions are asked: What is wanted? What is actually required? How is it used? In summary, the survey of requirements is intended to yield knowledge in the following areas: (a) information definitely required from each person at each level in order for him to fulfill his functions within the system; (b) reports required by management—their frequency and content; (c) statistics required about operations; (d) information at each level clearly required for making decisions; (e) actions resulting from such decisions; (f) relevant requirements; and (g) unnecessary requirements.

In order to evaluate and correlate the findings of the survey of requirements of each system and its components, findings should be systematically summarized for study. Preparation of the Summary Worksheet for the Survey of Requirements (Figure 5-4) is discussed in Chapter 5. The reports, information, and records purported to be needed to satisfy the requirements placed against the system will have to be evaluated as valid or invalid *outputs* in fulfilling the system's requirements.

chapter four

ANALYSIS PHASE OF THE SYSTEMS STUDY—CURRENT PROCEDURES

With the stated requirements of the system identified and recorded, the analyst or study staff is in a position to proceed with the three remaining portions of the analysis phase: the survey of current operating conditions (subsequently referred to as the "preliminary survey"), the survey of outputs, and the survey of inputs. Although each of these portions of the analysis phase is treated separately, as though autonomous, this is distinctly not the case. There is no alternative to performing the preliminary and the output and inputs surveys concurrently. The obviously interlocking character of the procedures applied in processing "inputs" to yield "outputs" satisfying the system's requirements calls for concurrent study of the three areas to arrive at a composite understanding and record of the system's operations. At the same time preliminary flow charts should be prepared in parallel with the understanding of the system's present procedures and be constantly revised until they picture the actual flow of work through the system. The technique of graphically representing the progression of work through a system is discussed in Chapter 6.

PRELIMINARY SURVEY—PREMISE

In the preliminary survey consisting of the analysis of current procedures everything about the system is closely scrutinized: its requirements, equipment, personnel, procedures, and their applicability to stated outputs of the system; the effect of other systems and functions of the library upon the system's procedures; and the flow of work through the system. The system should be analyzed in minute detail to make certain that all of its components together with their functions, decisions, and the actions taken, are isolated and precisely understood.

It cannot be overstressed that the assigned analyst should work *within* the system—become immersed in it—so he will actually know which of the system's problems cannot be analyzed in abstract fashion. He should gain acceptance by the system's personnel in order to learn the content of their jobs and how each staff member performs his assigned work. He should exercise strict objectivity in approaching the system without preconceived notions of the system's effectiveness, value, or need in the total library system, nor of the effectiveness, value, or need of any part of the system itself.

Although the entire analysis phase is concerned simply with observing and recording what exists and what is done presently, the analyst consciously identifies what does not exist and what is not done in the light of his understanding of the demands made of each system and his knowledge of the goals of the library. It is a cardinal principle at this stage that the analyst should make no suggestions of change in methods and procedures, however obvious the solution to a deficiency may be. Thus in the analysis phase preparation is made for the concluding major phases of a systems study—evaluation of the current system and design of a different system if required, where changes and adjustments felt to be necessary to improve present operations are incorporated. The continuation of the systems study to a successful conclusion depends entirely on the efficacy of the findings of the analysis phase.

PRELIMINARY SURVEY—WORKSHEETS

The tool of the preliminary survey is a methodical record of the present procedures being followed by a system. Preliminary survey worksheets are supplied for uniform approach and recording in the surveying of all systems under review. Figures 4-1a and 4-1b, 4-2a and 4-2b, and 4-3a and 4-3b are examples of filled-out preliminary survey worksheets.

WORKSHEET—GENERAL AND EQUIPMENT SURVEY

The first part of the worksheet (Figures 4-1a and 4-1b) identifies the system being studied, the name of the analyst, and the date of the survey. Here also a brief description of the system and its major activities is set down. To clarify the form's use two systems are given in illustration: an acquisition system of a central library (Figure 4-1a) and the cataloging system of a branch (Figure 4-1b). The described activities of the acquisitions system are to process book requests in accordance with selection policies, avoid duplication of library material, prepare book orders, check in books received, maintain accounting and statistical reports for all acquisitions. This is a digest of the chief activities of this system. The next item on the form calls for a statement of primary requirement, which in the case of the acquisitions system is plainly the ordering of library materials within budgetary provision. The manual of procedures, if available, is obtained at this time for comparative reference in the study of observed versus stated processes.

Turning to the branch library cataloging system (Figure 4-1b), we find it is briefly described as processing and indexing library materials for the following files: periodical index, vertical file of clipping materials, public card catalog, American Institute of Architects file, slides, prints and art works, and theses.

The primary requirement of the system is to provide timely and accurate updating of the library files and to process library materials for use as rapidly as possible.

A manual of procedures is available for this system and, if a valid manual, should facilitate the analyst's comprehension of the system.

The next portion of Figures 4-1a and 4-1b deals with the availability and use of equipment by a system for aiding its operations: typewriters, hard-copying machines, duplicators, adding machines, and similar common office machine aids. Special equipment may be typified by the keypunch, tape typewriter, punched-card reader, and card sorter. The questions to be answered in the equipment survey are exemplified by the following: (a) equipment used in present system; (b) other equipment available both in and outside the library; (3) location of each piece of equipment; (4) special features of any piece of equipment, such as tape typewriter with punched-card reader; (e) percentage use of each piece of equipment; (f) age and condition of each piece of equipment; and (g) authorization and procedure required for use of equipment outside of the library and its schedule of access, such as computer, centralized printing equipment, and services.

WORKSHEET—PERSONNEL

Worksheet Figures 4-2a and 4-2b continuing Figures 4-1a and 4-1b are designed for making a preliminary survey of positions and the incumbent per-

FIGURE 4-1a

PRELIMINARY SURVEY WORKSHEET (Rensselaer Libraries, Troy, New York)		
System: Acquisitions	Analyst:	Date:
Brief Description of System: To process requests from faculty and staff, eliminate duplication of material, prepare orders to publisher or vendor, check in books, maintain all accounting and statistical reports.		
Primary Requirement of the System: To order material to fulfill the teaching needs of the institute within the budget available.		
	Manual of Procedures Available Yes ☐ No ☒	

Equipment Survey				
Type of Equipment	Location	Special Features	Age/Condition of Equipment	Percent Utilization
Typewriters	Department	None	Replace on schedule	90
Photocopiers	Technical processing	Electrostatic, reproduces catalog cards	Service contract	75
Duplicator	Director's office	None	7 yr. good	Low
Tape typewriter	Technical processing	Punched card reader	3 yr. service contract	Low
Computer (System 360/50)	Computer laboratory	Upper/lower case print chain	Latest model	Low

Accessibility of Outside Equipment: Computer—available for administrative uses from 11 A.M.-1 P.M. and 4 P.M.-6 P.M. weekdays, and weekends.

FIGURE 4-1b

PRELIMINARY SURVEY WORKSHEET (Rensselaer Libraries, Troy, New York)		
System: Architecture Library—cataloging system	Analyst:	Date:

Brief Description of System: The system processes and indexes library materials in these files: periodical index, vertical file of clipping materials, card catalog, AIA files, slides, prints and art works, and theses

Primary Requirement of the System: To furnish timely and accurate input to to the above files and to process library materials as rapidly for use as possible.

Manual of procedures available Yes [X] No [ ]

Equipment Survey

Type of Equipment	Location	Special Features	Age/Condition of Equipment	Percent Utilization
4 manual typewriters	Office Workroom Circulation desk Unassigned	Card platen, extra wide carriage, library keyboard	1957—fair 1950—fair 1965—fair 1946—poor	80
1 electric typewriter	Office	Electric, regular and card platens, service contract	1961—good	0
LeRoy lettering set (spine lettering) vertical files, guide cards	Library workroom drafting table	Mechanical pen, mechanical lettering, miscellaneous templates, and points	1955—good	60 estimated
Dry mounting press (dry mounts photos, pictures for vertical file of clippings	Office	Will take oversize prints 15 × 18 in. will do 30 × 36 in.	1954—not working, need switch, good otherwise	90 estimated
Adding machine	Office	Electric, service contract	1964—very good	5 estimated
Electric eraser	Workroom cabinet	Smells	1959—excellent	100
Card master ditto for small 3 × 5 notices	Library workroom	Not used for catalog, could make catalog cards hand operated	1955—hard to use, expensive, messy	0

Accessibility of Outside Equipment: Xerox 720 in General Library, Technical Services Division; Photocopy machine in School of Architecture Office; Ditto machine in School of Architecture Office.

sonnel. This survey is an approach to two areas of long-standing interest to library managers. The first is the preparation of job descriptions to arrive at the content of the job and the skills required to do it properly. The second area is that of analyzing the qualifications of staff members to determine whether they are capable of and sufficiently trained for doing the particular work to which each is assigned. An attempt is made in the personnel survey to find out as completely as possible the qualifications required to perform properly a given job but without evaluating present job performance of the surveyed personnel.

The personnel survey is conducted through a combination of personnel and organizational records and personal interviews with each worker in the system. From organization charts, job descriptions, and personnel classification schedules,[1] the analyst seeks the following information: (a) the titles of the positions found in the system; (b) the job level or classification and grade of each position; (c) the special skills called for in each position; (d) the name of the incumbent of each position, noting vacancies or additional personnel not reflected in the organization chart or the budgetary authorization; (e) the actual job level—classification and grade—of the incumbent in a given job (is the incumbent working in or out of classification?); (f) the special skills possessed by the individual worker; and (g) the accuracy and up to dateness of job descriptions (if outdated or inadequate, modification of job analyses to the degree judged necessary by the analyst at this stage of the survey.)

This study of the attributes and deficiencies of personnel and the relationship between job requirements and the ability of job incumbents to meet those requirements is not an isolated exercise. In the course of the personnel survey the study staff or analyst has been accepting the prime responsibility for acquiring a knowledge of the major activities of the systems being studied. The analyst, further, in his interviews with personnel should have acquired sufficient awareness of details about elements of the jobs to grasp the interconnection between the major activities. Without this knowledge of the major activities and the details of the component operations, functions, decisions, and actions, the analyst is unable to proceed with the completion of the preliminary survey form or with the flow charting and analysis of the existing systems required in the next phase of the systems study:—evaluation of the competency of those systems.

[1] For guidance in the preparation and use of organization charts, job descriptions, and job classification schedules, see: *Office Management Handbook* ed. Harry L. Wylie, 2nd ed., Ronald Press, New York, 1958, Sec. 2, Organization of the Office, and Sec. 5, Salary Administration.

FIGURE 4-2a

PRELIMINARY SURVEY WORKSHEET PERSONNEL SURVEY					
Staff Positions	Job Level	Special Skills Required	Name of Person in This Position	Job Level	Special Skills Available
Acquisitions librarian	Senior librarian	Supervisory	Mrs. Jones	Senior librarian	Management education
Acquisitions assistant	Technical assistant	Bibliographic search	Mrs. Brown	Technical assistant	Language competency
Acquisitions clerk	Clerk specialist I	Typing	Miss White	Clerk specialist I	Trained key puncher
Acquisitions clerk	Clerk specialist I	Typing, bookkeeping	Miss Black	Clerk specialist I	Trained bookkeeper

FIGURE 4-2b

PRELIMINARY SURVEY WORKSHEET PERSONNEL SURVEY					
Staff Positions	Job Level	Special Skills Required	Name of Person in This Position	Job Level	Special Skills Available
Library clerk	Base	Typing, filing, dry mounting	Janet Rayfield	Base	French, college, minor in art with some architecture (BA, Sage)
Library clerk	Base	Typing, filing, dry mounting	Kim Worster	Base	background, secretarial skill, Vietnamese
Library clerk	Base	Typing, filing, dry mounting	Vacancy	Base	
Architectural librarian	Head supervisor	Knowledge of classification and cataloging system in use (Supervisory) ability	Virginia Bailey	Head supervisor	22 yr. experience in Architecture Library in May, knowledge of architecture literature, head since 1959—
Six student assistants	–	Must be a School of Architecture student, typing (some), lettering	–	–	knowledge of curricula

WORKSHEET—SYSTEM COMPONENTS

Figures 4-3a and 4-3b (System Components) serve to help the analyst organize and record his findings. The analyst in his gathering of details about how an operation is performed and the steps needed in a given operation supporting a subsystem's activities should have gained a generalized concept of the flow of work through a subsystem beginning with the first major activity triggered by an input into the overall system. The system components survey of each major activity or subsystem in work-step sequence furnishes a composite picture of the flow of work through the whole system consisting of its several subsystems. With the major activities and their component operations, functions, decisions, and actions recorded, the path of work through the system can be followed and pictorially represented in a flow chart (see Chapter 6). The analysis should be sufficiently detailed to enable the analyst to spot the same function occurring at two or more points within the same subsystem, within a related subsystem, or within an administratively separate system.

Figure 4-3a illustrates the analysis of a subsystem, the preorder search of the acquisitions system. The type of component is checked, which in this case is the subsystem preorder search. Its major requirements are described:—the avoidance of duplication of book orders and verification of all bibliographic data. The first operation, checked in the second column under Type of Component, is the searching of the library's records and consists of the operations of checking the official catalog, Library of Congress proof card file, and the library's public card catalog. The first function checked is the searching of the official catalog by title. The first decision, checked in the Decision—Action column, is that of finding out whether the book is in the library or on order. The action taken, checked in the fourth column, depends on the decision reached on the question, "Is the book in the library or on order: 'yes' or 'no'?" If "no" as illustrated, the chain of activity continues with the next step in the search procedure to verify the information needed for ordering and cataloging purposes.

The final decision that the information needed for ordering and cataloging purposes is complete ends the chain of activity in the preorder search subsystem and releases the validated book request to the order subsystem. There another chain of operations, functions, decisions, and actions are generated by this "output" of the preorder search subsystem.

In Figure 4-3b the preparation and maintenance of the periodical index file are treated as a subsystem of the branch library's cataloging

FIGURE 4-3a

PRELIMINARY SURVEY WORKSHEET
SYSTEM COMPONENTS SURVEY

In preparing the following analysis every effort should be made to list the components in the order that they normally occur in the present system.

Type of Component				Name of Component	Description	Standard Rate
Sub system	Operation	Function	Decision–Action			
X				Preorder search	Avoid duplication, verify all bibliographic information.	
	X			Search library files	Check official catalog, LC proof cards, public catalog.	6/hr.
		X		Search official catalog	Check official catalog by title.	18/hr.
			X	Decision	Is book in collection or on order?	
			X	Action	If yes, return request to originator.	
			X	Action	If no, continue search in LC proof card file.	
		X		Search LC proof card file	Check LC proof card file by title within year of publication.	18/hr.
			X	Decision	Is LC proof card available?	
			X	Action	If yes, attach to request and continue search in BIP.	
			X	Action	If no, order LC cards if card number is available and continue search in BIP.	
	X			Search trade bibliographies	Verify information.	5/hr.
		X		Search *Books in Print* (BIP)	Check BIP for bibliographic information for most domestic publishers.	16/hr.
			X	Decision	Is bibliographic information enough to order?	
			X	Action	If yes, forward to order subsystem.	
			X	Action	If no, search for bibliographic information in *Forthcoming Books.*	

PRELIMINARY SURVEY WORKSHEET
SYSTEM COMPONENTS SURVEY

In preparing the following analysis every effort should be made to list the components in the order that they normally occur in the present system.

Type of Component				Name of Component	Description	Standard Rate
Sub-system	Opera-tion	Function	Decision-Action			
X				Periodical index file	Maintain an index file to supplement the art index.	
	X			Reviewing journals	3 journals are regularly reviewed: *Architectural Record, Architectural Forum, Progressive Architecture*	
		X		Issue of *Architectural Record* obtained	This is a journal to be analyzed.	
			X	Decision	Has the issue been indexed according to indexing work record?	
			X	Action	If no, prepare draft slips for author and subject entries.	
			X	Action	If yes, send issue to vertical file preparation subsystem.	

system. The file is described as having the responsibility for indexing certain architectural journals to supplement the art index. The operation of reviewing these journals involves, as indicated, three regularly received journals. The first function checked is the receipt of an issue of the *Architectural Record*. This leads to a decision as to whether the issue has been indexed before or not, which necessitates a check of the work file. If the issue has not been indexed, the worker prepares draft slips for articles to be indexed. If it has been indexed, the issue is forwarded to the vertical file preparation subsystem.

It may be evident that these examples can be depicted in greater detail depending on the degree of detail desired or needed. For example, the "operation" of searching trade bibliographies to verify all bibliographic data necessary for maximum identification of a publication can be further detailed with respect to the sequential steps or functions that the searcher must go through in verification. This type of very detailed analysis is required for computer control of an operation in the absence of the heuristic capabilities exercised by the searcher in a manual system.

The matter of the degree of detail desired in the analysis of a manual operation can be illustrated by "typing a purchase order for a book." Within this operation are various functions, such as separating the manifold slips, filing one in the public catalog and one in the open-order file, mailing two copies to the vendor, and sending one copy to the card division of the Library of Congress as an order for printed catalog cards. This illustration also makes clear the distinguishing characteristics of an "operation" and a "function." It also shows that an operation usually consists of multiple functions or of sequential steps taken to complete the operation. Functions, then, are supporting parts or steps in effecting the consummation of an operation.

The finely drawn detailing of the functions, decisions, and actions described above is not particularly material to the broad picture of the system's operation that the analyst is looking for in the preliminary survey. The level of information as illustrated in Figures 4-3a and 4-3b is adequate at this stage of the systems study, however, for the analyst to prepare a generalized flow chart of the subsystem.

STANDARD RATE

We note in Figures 4-3a and 4-3b that the last column is headed "Standard Rate." It is used to record the number of units of output per unit of time in the performance of those operations and functions capable of such measurement. Although formal time-study techniques are applicable here, standard rates can be approximated with a high degree of validity simply by subjective observation. If it seems necessary to confirm the worker's rate of production a work sample can be taken. Records of production for a randomly selected hour daily for a period of a week will give a reasonably close approximation of standard rates of production or work output per man-hour. Anticipated daily or weekly rates then can be computed, setting quantity standards adjusted by a factor for worker fatigue, interruptions, and the making of decisions, to arrive at a reasonable day's or week's work.[2]

Determination of standard rates by work measurement, time study, work sampling, or by objective inquiry and observation clearly is needed for establishing work standards and recommending adequate staffing of an operation and a system. This determination is required not only in analyzing and understanding current procedures but in evaluating the current operating system, the second phase of the systems study (see Chapter 7) If, for example, the library's management places on the circulation system the requirement of shelving a minimum of 400 books per day, the capacity of the staff to meet or not meet this requirement must be known. When the number of books to be shelved and the number of books that presently can be shelved per unit of time is known, the system can be staffed accordingly. The grossness of this illustrative requirement is recognized, calling into play as it does a series of operations and functions of varying complexity, performed at varying rates of production or output. Thus the staffing of the entire circulation system must be analyzed for production "bottlenecks" in the flow of work and relief given them with additional staff or work simplification in order to meet the demand. Knowledge of the standard rates of all controlling functions permits identification of those functions likely to interrupt and retard the flow of work necessary for fulfilling the requirement placed on the system. A standard rate should be calculated for each function if it is measurable by units processed; if it is not, the calculation of the standard rate for the operation should consist of a composite of all functions performed in that operation.

QUESTIONNAIRES—JOB ANALYSIS AND JOB DESCRIPTION

Work standards with a satisfactory level of reliability can be set by a combination of job analyses and supporting job descriptions pre-

[2] Memo on Effective Labor Costs, in *Case Studies in Systems Analysis . . .*, ed. Barton R. Burkhalter, Scarecrow Press, Metuchen, N. J., 1968, pp. 9-10.

pared by the workers themselves. Figure 4-4 illustrates a job analysis form designed to allow for an exhaustive listing and indication of standard rates for the various functions performed by a worker. The predesignation on this form of all possible functions in a system will serve to remind the worker of functions that otherwise would be overlooked. The executed job analysis form is intended to provide the raw data for preparation of a more nearly complete description of the worker's job than otherwise could be predicted. This job analysis form filled out by all staff members, both clerical and professional, should yield a significant by-product for management:—revelation of duplicating and overlapping functions and of too much time being spent on "low level" work by professional and skilled staff.

A typical job description questionnaire is illustrated in Figure 4-5. The worker's summarization in this form of specific activities, distribution of his time devoted to such activities, his responsibilities, manner of performing his work, his working conditions, machines used, and so forth finds application not only in the analysis and understanding of current procedures but in the subsequent evaluation of the suitability of the current procedures as treated in Chapter 7.

The techniques and procedures of time study, random sampling, setting of work standards, standard rates or times, work simplification, and of job analysis through work sampling as related specifically to library tasks are fully treated by Dougherty and Heinritz.[3]

[3] Richard M. Dougherty and Fred J. Heinritz, *Scientific Management of Library Operations,* Scarecrow Press, New York, 1966. In addition, for a compact treatment of statistical sampling see Carl M. Drott, Random Sampling: a Tool for Library Research, *College and Research Libraries,* **30**:119-125, (March, 1969).

FIGURE 4-4

JOB ANALYSIS QUESTIONNAIRE

1. Complete the form for your position; then forward to department head.
2. Check each function you perform in your regular position; indicate the approximate time required for each item or the percent of your time devoted to that function.
3. Describe each item in detail in the Job Description Questionnaire.

Job Title:	Department:

Job Level:

☐ Secretary ☐ Clerk specialist I ☐ Clerk specialist III ☐ Librarian
☐ Student ☐ Clerk specialist II ☐ Technical assistant ☐ Other

		Standard Rate			Standard Rate
1	Accession books		21	Selecting and getting books for reclassification	
2	Alphabetize		22	Tying catalog cards together	
3	Alphabetize—catalog cards		23	Typing—authority file entries	
4	Original cataloging		24	Typing—cross references	
5	Corrections—LC books, catalog cards, and so forth		25	Typing—general	
6	Discussion (informal meetings)		26	Typing—LC cards	
7	Filing catalog cards —public catalog		27	Typing—labels	
8	Filing—shelf-list catalog		28	Typing—original catalog cards	
9	Insert book pockets		29	Photocopying—catalog cards	
10	Label books		30	Photocopying—LC proof slips	
11	Maintain authority file		31	Photocopying—other	
12	Property stamp books		32	Typing—book orders	
13	Maintain statistical records		33	Training new staff	
14	Meetings—staff and committee		34	Cataloging with LC copy	

FIGURE 4-5

JOB DESCRIPTION QUESTIONNAIRE
LIBRARY

1. Mr. / Mrs. / Miss — Last Name ____ First ____ Middle Initial ____

4. Department:

Cataloging

2. Job Title:

Clerk specialist—Cataloging

5. Name of Immediate Supervisor:

3. Regular Schedule of Hours of Work:

	from	*to*		*from*	*to*
Mon.	8:30	5:00	Fri.	8:30	5:00
Tues.	8:30	5:00	Sat.		
Wed.	8:30	5:00	Sun.		
Thurs.	8:30	5:00			

Length of lunch period or time allowed for meals:

1 hr (2—15 min breaks)

Total hours per week: 37½

6. Job Title of Immediate Supervisor

Head of Cataloging

7. Is Work:

Full-time x ?

Part-time ____?

Temporary ____?

If work is *temporary,* please indicate the part of the year you work:

8. Do you supervise others? Yes:____ No: x . If you do supervise others, list their names and titles. (If you supervise through subordinate supervisors, list the number supervised and their job titles.) Attach additional sheets if necessary.

a. I supervise *directly:*

Name	*Title*

b. I supervise through subordinate supervisors:

Number	*Title*

FIGURE 4-5 Continued

9. Please describe below in detail the work you do. Use your own words and make your description clear so that persons who are not familiar with your work can readily understand what you do. Use additional sheets if necessary. *Do not repeat duties.*		
Description	Work Performed (Description of Job Duties)	Average
Daily:		*Hours Day*
	Sort and Alphabetize LC cards	60 min
	Type LC number on proof cards (which are photocopied by another clerk, and *later* returned to the clerk w/the book for processing	20 min
	Odd jobs (corrections on LC cards: typing errors—pulling sets of LC cards from file and book from stacks; making changes on cards, such as entries, LC number, etc.	30 min
	Figuring out problems—searching for missing books, cards, etc.	38 min
	Changes on LC cards—erase and retype w/correct information (pagination, publisher, dates, series, added entries). Either have a new set of LC cards xeroxed from corrected card or correct whole set, then type (not every day)	20 min
	Filing—broken down into 3 different sections; consist of authors, where you have to pull the green slips from the drawer and match them with the white slips and author cards; titles, and subjects	90 min average, more or less depending on amount
	Sort LC cards by order number; pull greens from one drawer (then cards are matched w/books on the shelves. Latter is usually done by one of technical assistants	2 hr

FIGURE 4-5 Continued

Weekly:		*Hours week*
	Type cards—pockets for monographs and series. approx. 4 hr day	20 hr
	Glue in pockets on glue machine, usually for monographs and series	1 hr
	(if Mary Ann is absent, I glue in all pockets)	2 hr
	Pull monographs and series from stacks—type labels, and accession cards and book (approx 4 hr/day)	20 hr
	Type LC cards and pockets for new books	5-6 hr
	Pull Dewey cards and books for reclassification	2-3 hr
Monthly:		*Hours month*
Occasional:		*Hours period*
(indicate period—weekly, monthly, quarterly, annually.)	Use Se-lin-type writer (for typing labels)	3-4 hr
	Xerox machine	5 min
	Type cards for original cataloging	1 hr
	Type cross-reference cards for the file	1-2 hr
	Train new clerk; one every 6 mo. to a year	4-5 hr day
	Social committee—buy coffee, tea, birthday cards, etc.	1 time every 1½ yr
	Weekly social committee—make coffee, tea, clean up lounge	1 time every 5 or 6 mo (30 min/day; 1 week)

FIGURE 4-5 Continued

10. List machines and equipment used and indicate about how much of your working time you spend on each:

Machine	Percent of Time	Machine	Percent of Time
Glue machine	5		
Typewriter	50		
Eraser	3		
Se-lin	2		
Iron	2		

11. Describe the type of instructions you receive before you start your various job tasks. Also explain how much and what kind of guidance you receive from your supervisors, manuals, or established procedures while you are working.

We receive verbal instructions for any new jobs that we may do and if we have a question, it is answered by our supervisor.

12. Explain in what way and how often your work is reviewed by your supervisor.

Catalogers revise our daily filing.

Books and LC cards are revised by a technical assistant.

Monographs and series—problems are worked out mostly with a cataloger.

Sorting and filing new LC cards for a technical assistant.

13. In performing your job what contacts do you make with departments other than your own, with outside organizations, and with the general public.

Searching for missing books or books on reserve—go through Circulation.

Other than a question or two in reference, book order, serials, I do not work too much with them.

14. Usual work conditions, if applicable:

x Hot
x Cold
___ Sudden temperature changes
___ Humid
___ Dry
___ Wet
x Dusty
x Dirty
x Noisy
x Cramped quarters
x Working with others
x Working around others
___ Working alone
___ Other (describe any conditions not listed) ______________

____________ Date

____________ Employee's Signature

FIGURE 4-5 Continued

15. *Statement of Immediate Supervisor:* Please comment on statements of employee. Indicate any exceptions or additions and what you consider the most important duties of this job.

Most important duties: reclassification of volumes from Serials into the LC book collection.

Date	Supervisor's Signature
Date	Department Head

chapter five

ANALYSIS PHASE OF THE SYSTEMS STUDY—DETERMINATION AND SURVEY OF INPUTS/OUTPUTS

The third and fourth distinct but related portions of the analysis phase of a systems study are those of determining and analyzing the "inputs" and "outputs" of the system. Input means just what it says—information that is a matter of record within a system. Output also means what it says—information transmitted from the system. The flow-chart symbol for an input or output is the same.

If confusion arises in distinguishing input/output from requirement or demand, it is helpful to remember that a requirement or a demand is a task assigned to or assumed by a system or subsystem to satisfy a segment of the overall goals of the total system. The input/output of the system results in the performance of the task. This performance is accomplished by means of information at hand that can be sorted to fulfill the requirement or the demand and the result of this sorting is output.

In this chapter the term "input/output" is used because of its duality of purpose. An output of one system or subsystem frequently becomes an input of another and it is the complex chain set up that needs to be analyzed. One input—for example, a book request from a library user—sets up a chain that is obvious as we follow this request from acquisitions through cataloging to circulation. Subsystems of each of these systems are involved and it is in the survey that the picture takes shape.

Each input/output should be followed through the system to its destination, finding out in its course the operations utilizing the input/output, the information added to it if this occurs, and how it is acted on to meet its primary purpose of contributing to the satisfaction of a stated requirement of the system. The questions to be answered are (a) is the input/output needed to meet the requirement and (b) is the input/output information adequate with respect to content and form to permit efficient satisfaction of the requirement?

INPUT/OUTPUT WORKSHEETS

In following either a specific input or output through the system the analyst should make use of an executed Worksheet for the Survey of Inputs (Figures 5-1a and 5-1b) covering the handling of each input or a Worksheet for the Survey of Outputs (Figures 5-2a and 5-2b) covering the handling of each output at each "level" as it progresses through the system, resulting in multiple records of the same input/output as it goes through transformation, modification, and change to achieve its objective. The term "level" can be equated to the individual worker who performs certain functions, makes decisions, and takes action in his manipulation of the input/output for referral to a fellow worker or to another system. Here the analyst should make certain that the level at which an input, for example, enters the system is the level where it is used or the action is taken contributing to the processing of the input. It is not uncommon for a worker to receive data or requests for no other reason than that of referring them to someone else in the system who does the work the input calls for. Such interim handling, functionally wasteful, clearly delays the processing of the input and slows the progress of the system's work in meeting its requirements. Hence the level at which the input entered the system is not proper and appropriate revision is necessary at the design stage of the study.

The handling of a book invoice exemplifies the processing of an input/output at different levels. The invoice is received by the package receiving clerk. He uses it in confirming that the items listed are in hand or that the invoice is incorrect in this respect. The invoice is either referred to the accounts clerk for recording and clearance for payment or it is referred to the acquisitions clerk for claiming the missing item.

The three workers in this example each represent a level and from each a worksheet is obtained indicating the functions, decisions, and actions taken in processing one basic input/output—a book invoice.

SOURCE AND TYPE OF INPUTS/OUTPUTS

As in the case of the requirements of a system the inputs/outputs of a system are received from many sources and in some instances the form and content of the input/output cannot be altered or controlled, however essential to improved operating efficiency and best results a change might be. The analyst should become cognizant of all sources of inputs/outputs to know whether or not opportunities exist for proposing redesign of the form and content of given inputs/outputs in the interests of efficiency, economy, and effectiveness. The sources may be (a) from outside the governing administration of the library; (b) from outside the library but within the governing administration; (c) from within the library; and (d) from within the library's system being surveyed.

With the inputs/outputs of a system being derived from the requirements placed on it the determination and survey of requirements unavoidably bring attention to the inputs/outputs of the system. This again illustrates the parallel, interconnecting character of the systems study in which information is gathered simultaneously in related areas to yield an interlocking picture of the system's activities.

FORM OF INPUT/OUTPUT

The form of the input/output of a system has bearing on the efficiency and economy of operations. Is it handwritten in an illogical fashion necessitating typing and rearrangement and addition of information before action can result from or be taken on the input/output? If it is a typed or photocopied request, does it have to be recast for processing in the established routine of the system? One of two courses is open: effect a change in the originator's method of producing the input/output including arrangement of the content of the form or adapt the system's procedures to the input/output in whatever form received. The latter course usually can only result in continuation of operating problems and inefficiency.

For example, the book purchase request brings into play two factors: control of the source and content of the input/output and the design of the form of the input/output. To illustrate, book purchase requests may be submitted in the form of handwritten scraps of paper, clippings, marked book review pages, and listings and may often lack critical elements of book identification. Such a situation materially slows the acquisition process, reduces personnel efficiency, and increases operating costs per title processed. To overcome these undesirable, inefficient, and uneconomical operating conditions a printed request form may be designed with the items of required information arranged in the format that is best for both the preorder search functions and the preparation of book orders. Control over the source of this input/output, the requestor, may be enforced by the library's not accepting the book requests unless each is submitted on the newly designed form. Because preorder search personnel would be able to use this input/output directly without the former recasting and typing of the request, this should improve their rate of output to the order subsystem where the order could now be typed directly from the original input/output.

This manner of adjustment is confined to sources of input/output within the province of the library. Libraries, in general, may lack the leverage to persuade sources outside their governing administration to change the form of their inputs/outputs for best handling within the library's operational design. Libraries with extensive acquisitions' budgets, however, should be able to ask and receive the acquiescence of contracting vendors to use a suggested form of invoice and bibliographic data best fitting that system's design. If machine-produced records in the form of punched cards, paper tape, and magnetic tape are received by a library employing computer controls, these forms of input/output should be evaluated with respect to direct incorporation into the library's computer system to avoid the additional step and expense of regenerating such records.

The question of control or influence over the source of such machine-produced records may be further exemplified. If an input/output comes from the comptroller of the governing administration of the library, it is conceivable that discussion with this local officer will result in the form and content of his input/output being modified or changed to a more useful form for the library. If the library is accustomed to issuing invoices of charges made in its photocopying service, for collection by the comptroller, any change in the format of the library's invoice should be discussed with the comptroller. If the invoice being used requires a variety of signatures and other information manually added to this input/output, then a new invoice form utilizing automated billing techniques may be designed to take the currently accepted procedure into account or if viewed as an improvement, the requirements of the current procedure may be modified.

As another illustration the monthly statement of expenditures for library materials from the comptroller may be of minimal control significance to the accounting and reporting subsystem of acquisitions because it lacks indication of the encumbrances standing against the book funds. The elimination of this input/output and the assumption by the acquisitions system of up-to-date recording of expenditures and encumbrances might be more meaningful to both the library and comptroller's office which could be supplied this record at any time desired.

Another input/output considered is the "verbal communication" type. This form of input/output, especially in the fulfilling of requirements, should be eliminated to the fullest extent possible unless it can be reduced to recorded form. The proliferation of verbal inputs/outputs unrecorded at their entry into or exit from the system will likely result in repetitive handling of inquiries and unnecessary disruption of the planned flow of work. Should urgency or need for a book by a user require telephoning the vendor or publisher for a report on the delay of receipt of the item, the information obtained should be written on the open-order record. There is extensive verbal communication in reference work with many inquiries and answers that should be recorded. Wasteful repetition of the same search process at a later time may well be avoided if records are made and kept of these verbal inputs/outputs.

The variant types of inputs/outputs are (a) manual—for example, handwritten or typed information to be acted on, clippings, lists, book charges, overdue notices; (b) machine record—such as punched card, circulation transaction card, magnetic tape (LC MARC, other machine-readable data); and (c) verbal communication of information to be acted on—such as renewal of book charges by telephone.

SURVEY OF INPUTS/OUTPUTS—ANALYTICAL PRINCIPLES

Some indication of the procedure to be followed in the survey of inputs/outputs was given in the beginning of this chapter. The requirements are (a) the isolation of each input/output to a system, including each variant of the same input/output as it is used at each functional level; (b) identification of the level at which each input/output enters or leaves the system; and (c) determination of the need, use, and path of each input/output as it progresses through the system to its ultimate objective and disposal.

In order to satisfy the requirements of the detailed analysis in the input/output survey numerous questions about each input/output must be posed and answered.

1. What is the form of the input/output? Is it necessary that the form be changed for use within the system? Keeping in mind the matter of control over the input/output, the analyst is mainly concerned with the processing needed to use the information supplied. Does the vendor's invoice have to be rewritten or recast for use? Do requests for books have to be rewritten or recast for efficiency in preorder search operations? Do other systems in the library generate requisitions for books in incomplete or variant form? These and similar situations are investigated for unnecessary data processing that can be eliminated to yield lower unit costs of operation and more purposeful utilization of personnel. The underlying question is can the format of the input/output be changed if needed to improve operations?

2. Where and how does the input/output originate? This question strikes at the critical problem of control over an input by the system. The design of a system or improved data processing methods and procedures is significantly influenced by the extent of control over the form and content of the inputs/outputs. An appreciable lack of control over key inputs/outputs can prevent the design of the most economical system, or the most effective or the most efficient system, or the design of a system possessing a combination of these attributes. This can be illustrated by the library that has applied the Dewey decimal classification through its several editions and has decided to continue with it in any event. Having to develop this classification data for each title and to reclassify and recatalog material previously classified under earlier editions in order to maintain collection integrity may yield the most effective results but certainly not with the maximum possible economy and perhaps not with any degree of efficiency as measured in terms of units completed per unit of time against a cataloging operation using the Library of Congress classification scheme without modifications.

3. At what level does the input/output enter or leave the system? At what levels is it used? These are leading questions with the objective of making certain that the level at which the input/output actually enters or leaves the system is the level at which it is first or last actually used or acted on. As mentioned earlier it is not uncommon to find improper referral or interception of incoming data at a level where the worker takes no more action than that of passing on the input/output to another level that may or may not be the proper level for use of the input/output. This illustrates the principle in the survey of inputs/outputs of establishing the path of each input/output from its entry through its progression and exit, from level to level or from worker to worker in the system.

4. What information from the input/output record is required by

each level receiving it? This question keeps emphasis on the isolation of the needless entry and interception of an input/output at a given level or levels and at the same time seeks identification of the precise portion of the data of an input/output used or acted on at each level where the input/output is found. For example, a reference question about the library's annual review series holdings is properly addressed to a reference librarian at the information desk. The librarian may either check the serials catalog or, this failing, refer the user to the serials system for more up-to-date information. In systems analysis, however, it could be discovered that such reference questions (inputs) are being asked inadvertently of a student assistant who mistakenly is referring users to the acquisitions system which refers the users to the serials system which, in turn, refers the users to the information desk, the normal entry point of these inputs.

5. Is information added to the input/output record for use at other levels or is a new input/output record created? The point being investigated is the actual need for adding information or for recasting the input/output in acceptable form in relation to the efficient use of the system's staff.

An example of a book invoice as an input/output illustrates the adding of the number of the library account to be charged, unquestionably needed at the accounts posting level. *Recasting* or creation of a new input/output record is typified by the book purchase request submitted in a form requiring it to be rewritten in standardized format for efficient processing by the preorder search subsystem. This problem may be further typified by the "class reserve" reading list in an incomplete, inaccurate, and confused format received by the circulation system which is required to correct and recast it in the standard form needed for efficient processing.

6. What functions, decisions, and actions are triggered by the input/output? What workers are involved and what does each do in processing a given input/output through the system?

For example, the "book" as an input has numerous levels of workers logically concerned with the processing of this one type of input. A Worksheet for the Survey of Inputs (Figures 5-1a and 5-1b) must be filled out for each manipulation of the book as an input for securing a correlative view of the functions, decisions, and actions entering into all aspects of processing it.

7. What is the final disposition of the input/output? What files and records does the input/output affect? Are there written procedures describing the functions in processing the input/output? The processed input/output may be disposed of by (a) filing, (b) recording pertinent

information from it and discarding the original input/output, (c) being discarded entirely, and (d) being transmitted as information for further action or manipulation at another level. The effect of an input/output on files and records is illustrated by the book invoice that has a direct bearing on the updating of the book funds' accounting records. The book purchase request may be discarded after being transformed into a purchase order, one part of which may become a record of book on order in the public card catalog. Another example of disposition of inputs/outputs is in a computer-based system where data are entered on punched cards that subsequently are discarded.

TYPES OF INPUTS

In the interests of an orderly and more detailed analysis and survey of inputs that vary in their importance to and effect on the operations of a system, inputs may be logically classified into four types: primary, functional, instructional, and informational.

1. *Primary Input.* This type of input is characterized by its capacity to activate a major activity—a subsystem of a system. It usually calls for activity entailing major data processing functions, decisions, and actions. In the acquisitions system an approved book purchase request, a primary input, "triggers" the preorder search subsystem into activity; receipt of a book activates the receiving and checking subsystem. Similarly Library of Congress cataloging data activates the card preparation subsystem of the cataloging system. Another primary input, an interlibrary loan request, activates the interlibrary loan subsystem of the circulation system. A final example of a primary input is a user's request for a literature search by the reference system.

2. *Functional Input.* This type of input is information or inquiries involving one step—a minor clerical function requiring only routine action. The functional input is further characterized by being acted on by the worker at the level of entry in the system and usually goes no further. For example, notice of delayed publication only requires the clerk receiving the notice to post routinely the date of expected receipt to the open-order record. A further illustration of the functional input is the verbal input, "Has the book arrived in the library?" A routine check of records and the giving of a "yes" or "no" answer is the completed function, obviously not going beyond the level of entry or the worker receiving the inquiry. Another example of the functional input lies in the acceptance of Library of Congress printed catalog cards without revision. The function simply is that of typing the added entries

as found on the unit card and requires no decision making. A last example is the telephone request(input) to renew a book. The book charge may be simply redated and the charge refiled.

3. *Instructional Input.* This type of input is one that modifies or explains a primary input by providing instructions for the handling or disposition of the primary input. Such instructions can come from within the responsible system as well as from sources outside of it. For example, a cataloging worksheet gives directions for the processing of books, the primary inputs to the cataloging system. A professor requests the purchase of a given book and at the same time requests that it be placed on reserve for his class when it arrives. This then is an instructional input from outside of the acquisitions system specifying variant handling of one of the system's primary inputs, the book. A further illustration is an instruction that journals shall not be sent on interlibrary loan thereby modifying a primary input of the interlibrary loan subsystem, the off-site request for library materials.

4. *Informational Input.* This type of input is defined as requested or unsolicited information that may be transformed into any of the other three types of inputs depending on the action taken on that information. An announcement of a new title is on the one hand useless information if the book is regarded as not fitting the library's need. On the other hand the decision to order the book makes the announcement a primary input to the preorder search subsystem. To illustrate solicited information the library subscribes to the Library of Congress proof-card service. These cards, filed for possible future use in title verification and card reproduction, represent informational inputs. A slip withdrawn from the file becomes a primary input of the card preparation subsystem of the cataloging system. If cataloging operations are computerized, the slip may become a primary input for preparing the computer record. As a final illustration a new filing code may be suggested for the library's catalog records. If disregarded, this new code is simply information; if it is adopted or adapted by the library, it becomes an instructional input.

For the analyst the first two types, the primary and functional inputs, simply must be identified. However, the need for the next two types of input, instructional and informational, should be analyzed. The instructional input should be minimized by setting up procedures to cover all possible nonroutine functions. Every effort should be made to eliminate the instructional input by developing, where possible, standard procedures for nonroutine functions. Regarding the informational input, it is essential for the analyst to identify the decisions and actions occurring both within and outside the system responsible for the entry of the informational input into the system. Further, he should identify the highest level at which these decisions and actions originate and occur. Finally, the analyst should decide which of the other types of input (primary, functional, and instructional) the informational input will become if accepted.

SURVEY OF INPUTS—WORKSHEET

The foregoing questions needing to be answered about each specific input at each level are clarified in the executed Worksheet for the Survey of Inputs (Figures 5-1a and 5-1b) which illustrates the Library of Congress proof card as the input to the acquisitions system.

It is noted that the proof card is designated as an informational type of input simply because it is available for use if needed. Nothing is done to change its form or content nor is it put to any use except to be added to a file of proof cards. Being an informational input it can be made viable only if it becomes any one of the other three types of input. The Library of Congress proof card obviously may become a primary input to subsystems of both the acquisitions and cataloging systems. It becomes a primary input to the preorder search subsystem the moment it can be matched with a book request. It becomes a primary input to the card preparation subsystem of the cataloging system on receipt of the book represented by the proof card. The answer of "no other" to the question on the worksheet, "Which of the other types of input does it become?" is true at this level of the processing of the proof card.

As a primary input to the preorder search subsystem it will be used in its original form and in the card preparation subsystem it also will be used in its original form as the basis of the library's permanent catalog records. It is also seen that the decisions, actions, and functions are of an elementary and limited nature in this operation with the end product being a file of the inputs unchanged in any respect.

Figure 5-1b is a filled-out example of the second part of the Worksheet for the Survey of Inputs continuing with the proof card as the input to the acquisitions system. This second part is directed in general to questions of the utilization of personnel in this operation: can the job be done more efficiently to conserve worker time for other purposes? Can procedures be changed to effect a time saving? Can the volume of units received, or processed, or both, be reduced? Does the end product serve a purpose justifying the time taken to provide it? Time and motion and use studies may be required for definitive answers but initially considerable reliance can be placed on the interviewee's estimates of time taken in processing the input and the observed fre-

FIGURE 5-1a

WORKSHEET FOR THE SURVEY OF INPUTS (PART 1)

1. Prepare a copy of this worksheet for each input received.

2. Attach completed sample of the *input.*	Name of Input: LC proof card	
System: Acquisitions	Analyst:	Date:
Name of Person Interviewed:	Position: Senior acquisitions assistant	
Name of Supervisor:	Position:	

Review the description of the types of inputs given below and check the one that best describes this input.

☐ *Primary Input:* Involves major clerical, professional, or mechanical functions, decisions, and actions.

☐ *Functional Input:* Involves only minor clerical functions requiring no decisions.

☐ *Instructional Input:* Modifies or explains a primary or functional input.

☒ *Informational Input:* Potentially may become the basis for one of the other types of inputs. Please answer additional questions below on informational inputs.

If informational input, who makes the decision that it is to enter the system: Respondent

If *you* make the decision, on what is it based: Written procedures to discard certain classifications, juvenile titles, and nonbook forms.

Which of the other types of input does it become: No other

Source of Input: Library of Congress	Do you have authority over this source: ☐ Yes ☒ No

Form of Input: ☒ Printed ☐ Typed ☐ Punched cards ☐ Paper tape ☐ Verbal input ☐ Other (describe)

Is input used in it's original form: ☒ Yes ☐ No	Does the input enter the system at your operation: ☒ Yes ☐ No

Describe decisions you make in processing this input: Is proof card for a serial? Is proof card for an architecture title? Should proof card be discarded? Should proof card be filed in proof card file (according to selection policy)?

Describe what action you may take as a result of your decisions: Discard designated proof cards. Forward to serials librarian. Forward to architecture librarian. File in proof card file.

Describe the functions you perform in processing this input: Arrange for filing.

FIGURE 5-1b

<table>
<tr><td colspan="6">WORKSHEET FOR THE SURVEY OF INPUTS (PART 2)</td></tr>
<tr><td colspan="3">How many do you receive:
_____ Daily 200 Weekly _____ Monthly</td><td colspan="3">How many do you process:
_____ Daily 200 Weekly _____ Monthly</td></tr>
<tr><td colspan="3">How much time do you require to process this input: 45 min.</td><td colspan="3">What is the disposition of this input:
☒ destroy ☒ forward ☒ file</td></tr>
<tr><td colspan="3">If used by other functions to which do you send it: Architectural librarian, Serials librarian</td><td colspan="3">If verbal input do you record for later use by you or others: Not apply</td></tr>
<tr><td colspan="3">If you do record verbal input, in what form: Not apply</td><td colspan="3">What is the name of the form: LC proof card</td></tr>
<tr><td colspan="2">How long is this input kept in active files: Indefinite</td><td colspan="2">How long kept in inactive file: None</td><td colspan="2">How often is this file referred to: Daily _____
Active _____ Inactive _____</td></tr>
<tr><td colspan="3">How many of these forms are filed:
_____ daily 100 weekly _____ monthly</td><td colspan="3">Are written procedures available ☒ Yes
for the processing and control ☐ No</td></tr>
<tr><td colspan="6">What information do you add to this input? Where does it come from? If other form, name it. Circle this information in red on sample:
None</td></tr>
<tr><td colspan="6">Do you copy any information from this form on to other forms or records; if so, why? Circle this information in green on sample:
None</td></tr>
<tr><td colspan="6">Is any of the information on this form unnecessary for your operation? Circle this information in blue on sample:
Most of the information, excepting LC card number used for filing in proof card file.</td></tr>
<tr><td colspan="6">Does any of the information on this form appear on other forms used in this system other than those mentioned above? Circle this information in black:
No</td></tr>
<tr><td colspan="6">Use this area for your comments: Please do not forget to attach a completed form:
Should be selective about classifications received by proof card subscription; filing time exorbitant in relation to number of proof cards used from file; LC should put shipments in classified order; or by LC card number; keep "continued" cards with their first cards.</td></tr>
<tr><td colspan="6">This area for the analyst's comments:
Method needed for purging proof card file of unneeded information after x years? Arrange file by year digits of the LC card number? Can processing time be reduced?</td></tr>
</table>

quency of consultation and the extent of the usefulness of the proof-card file. Further clues about the proper or wasteful use of worker time can be gathered from the interviewee's comments called for near the end of this part of the worksheet. In the filled-in comments it is seen that internal policy change is implied with respect to the subscription for proof cards; it further is thought that the maintenance of the proof-card file is "not worth the candle"; and in connection with the ever-present problem of authority over the source of the input change in methods and procedures of the card division of the Library of Congress, well outside of the venue of the library, is suggested.

SURVEY OF OUTPUTS—WORKSHEET

The Worksheet for the Survey of Outputs, (Figures 5-2a and 5-2b serves to identify an output by name and the job level at which it originates in the system and its type is explored. Is the output a report, financial or statistical? Does the output result in a record or file maintained by the worker? Is it a statistical, financial, or bibliographic record? Is the output a form, such as a purchase requisition, a notification of book received, or an overdue book notice?

The requirement or requirements for which given outputs are generated should be described. Referring to the sample Worksheet for the Survey of Requirements (Figure 3-1), a requirement of the accounting and reporting subsystem is the maintenance of accurate and current balances for all book fund accounts in order to generate the output that is a tabulated fiscal report on the status of each account as of the end of each month.

In the survey of outputs, as in the determination and survey of the requirements of a system (Chapter 3), the worker being interviewed is asked to describe the functions, decisions, and actions that enter into the production of each output (the tabulated fiscal report in this instance). The functions include the updating of the posting sheets for each book fund to the month's end, preparation of a draft copy of the required consolidated report, verification of the accuracy of the draft report, and reproduction of the finished report in the number of copies needed for distribution.

The decisions required in preparing this output include determination of the cut-off date in posting to individual account sheets in order to release a tabulated report on the first work day of the succeeding month, determination that the balance for each account is accurate, and determination that posting of new transactions after the cut-off time will not occur until the tabulated report is completed.

The actions taken completing this output's cycle are the correction of accounting errors, typing of the report, the distribution of the copies of the report to designated persons both within and outside the library, and the filing of one copy in the monthly report file maintained in the accounting and reporting subsystem.

Attention again is drawn to the subtlety of the difference between requirements and outputs. The requirement of maintaining accurate and current balances is simply a designated objective. The need for transmittal of information or preparation of a report activates and justifies that objective or requirement of the operation. In analyzing the requirements of a system an attempt is made to isolate the outputs that satisfy each requirement; conversely, in analyzing the outputs of a system the specific requirements that the outputs are supposed to satisfy should be identified.

The second part of the Worksheet for the Survey of Outputs, Figure 5-2b, is designed to elicit evaluative information about the records and reports made and maintained as the basis of the output under investigation. Because the opening questions about the record kept and the report made are of a quantitative nature, they are self-explanatory and do not particularly require illustration. Answers to many of the questions here serve to prepare the analyst for the evaluation of current methods and procedures. He should be able to determine the necessity for certain records and reports as measured by the extent and frequency of use, importance of the purpose served, and by their apparent duplication and ensuing uselessness. Underlying most of these questions is the possibility of better utilization of staff through alleviation of unimportant functions and unnecessary proliferation of records and multiple copies of reports for distribution to points where they may serve no useful purpose. What is also being analyzed is the availability or unavailability of properly designed forms permitting preparation in one writing of multiple copies of a basic record that is needed for various purposes. The multiple-part book purchase form is a case in point; here the various parts carrying the same information produced in one machine operation are required for full control of the book order and processing of the book when received. Another example is a multiple-part loan transaction card where extra parts are used for overdue notices.

The remainder of the questions are for the purpose of finding out where the output record or report originates. If it does not originate with the interviewee and is received by him for adding information, where does it come from and what information does he furnish toward completion of the output? Is his function simply that of entering a bit of factual data or that of manipulating or interpreting the data or figures

FIGURE 5-2a

WORKSHEET FOR THE SURVEY OF OUTPUTS (PART 1)		
1. Prepare a worksheet for each output that is prepared or maintained. 2. Attach a completed copy of each output (record, report).		
System: Accounting and Acquisitions—Reporting Subsystem	Analyst:	Date:
Name of Person Interviewed:	Position: Acquisitions Clerk I	
Name of Supervisor:	Position: Acquisitions librarian	
Name of Output: Monthly financial report		
Type of Output: Bibliographic ______ Inventory ______ Fiscal ___x___ Statistical ______	Record ___x___ Report ______	
Identify the requirement(s) that use the output: Maintenance of accurate and current balances of all book funds.		
Describe the functions you perform in preparing or maintaining this output: Update posting sheets for each book fund to month's end, preparation of a draft copy of the required report, verification of its accuracy, and reproduction of the report for distribution.		
Describe the decisions: Determine the cut-off date for posting, making sure that the balance for each account is accurate, and make decision not to include any new transactions after cut-off date.		
Describe the action taken: Correct inaccuracies, type report, distribute completed report, file copy in monthly report file.		

FIGURE 5-2b

WORKSHEET FOR THE SURVEY OF OUTPUTS (PART 2)

Records:	Present size of file:	How many records added to the file: ___Daily ___Weekly ___Monthly

Type of File:	How file arranged:

Are written rules available: yes ☐ no ☐	How long is record kept in active file:	How long kept in inactive file:

How often is file referred to:	By whom:	Other disposition of impermanent records: ___Transfer ___Destroy
How much time required to prepare record:	How much time spent on filing:	

How much time spent in updating:

Reports:	How many copies:	What use is made of each copy and by whom:

Report Status: ___Intermediate ___Final	Method of Preparation:

How much time spent in preparing report:

What information, if any, is added to the output and where does the added information originate. Underline this information in red on sample and state its source below:

Is any of the information in this output transcribed to other forms or records. Underline this information in blue on sample and identify the other forms and records to which it is transcribed:

Does the information contained in the output appear on other records independently originated in or outside of the system. Underline this information in black on the sample and identify the other records where the information also appears:

Use this area for your comments; please do not forget to attach a completed form:

This area for the analyst's comments:

carried by the record received by him? Answers to inquiries of this nature are pertinent because the completion of a report or record through addition or manipulation of data at two or more levels within a system may be more the rule than the exception.

The remaining two questions in the survey of outputs strike at the common problem of duplication of effort. Proceeding on the principle that a record once made should not be duplicated but should be transmitted as a completed function at all the operating levels through which it logically flows, the analyst is constantly on the alert for evidence of multiple transcription and duplication of the same information in other records originating independently elsewhere within the system or by an associated system in the library.

Whether the output is a record or a report, the analyst's study of an activity and a system is aimed at gaining complete understanding of the information furnished by each output, gaining knowledge of the sources of such information including the information needed, if any, from other activities or systems to complete the output, and recording how many outputs the worker prepares or maintains, or both, and the amount of the worker's time consumed in this work.

These and other factors entering into the survey of outputs are now outlined, keyed to the instructions and questions in the Worksheet for the Survey of Outputs (Figures 5-2a and 5-2b).

1. Name of output and job level at which it originates in the system with the book receiving clerk, the accounts clerk, the acquisitions assistant, the circulation assistant, and so forth.

2. Type of output:—bibliographic, fiscal, statistical, and inventory—and the form it takes—sheet or file record or report.

3. The requirements of the activity satisfied in part or in whole by the output.

4. The functions of the activity giving rise to the output.

5. The decisions needed to be made in preparing the output.

6. The actions resulting from these decisions; effect of such actions on the output.

7. Determination for each record maintained by the system of the (a) size of the file and the number of records added periodically; (b) type of file (for example, ledgers, invoice files, card records) and its arrangement. (c) availability of rules for file preparation and maintenance; (d) length of time records are kept in active file, inactive file, and other disposition—for example, destroy or transfer; (e) frequency of reference to the record and by whom; (f) time used to prepare and maintain record; and (g) time spent updating record and its file.

8. Determination for each report prepared by the system of the (a)

number of copies prepared; (b) recipients of the report; (c) report status, intermediate or final; (d) method of preparation; and (e) time used in preparing report.

A sample copy of each output is obtained from each person responsible for preparing it together with answers to the following concluding questions of the worksheet:

1. What information, if any, is added to the output and where does the added information originate?

2. Is any of the information in this output transcribed to other forms or records?

3. Does the information contained in the output appear in other records or reports independently prepared and maintained by another activity in the system or in a system outside of the library?

SURVEY OF INPUTS—SUMMARY WORKSHEET

The culminating step, as in the case of requirements and their associated outputs, is the systematic summarizing of the data gathered about each input at each level where it appears using a Summary Worksheet for the Survey of Inputs (Figure 5-3). The use of the Library of Congress proof card as an example is continued, together with the book request and the book order. Understandable abbreviations or accepted codes have to be used for most of the entries called for in the form. This "shorthand" method is used to obtain a consolidated and generalized overview of the inputs to a system, their multiple handling at the various levels, and their ultimate disposition as indicated in the individual input survey sheets resulting from the input analysis at each job level in the system.

These summarized data coupled with the marked copies of samples of input documents used at each level in the system permit identification of the characteristics and content of inputs in relation to the processing steps each goes through.

SURVEY OF REQUIREMENTS (OUTPUTS)—SUMMARY WORKSHEET

The Summary Worksheet for the Survey of Requirements, Figure 5-4 is used in summarizing the survey of the outputs of a system. The summary worksheets should have attached to them the illustrative records

SUMMARY WORKSHEET FOR THE SURVEY OF INPUTS

System: Acquisitions | Analyst: | Date:

Name or Description of Input	Type of Input	Form of Input	Deposition of Input	Number Processed D, W, M	Originating Source	Input Pt. of Entry to System	What Operations Receive this Input	What Operations Require this Input
LC proof card	I	Print	Filed	200W	LC	Acquisition assistant	Proof card file	Proof card file
LC proof card	P	Print	Put with book request	20D	Proof card files	Searcher	Order	Card prep-aration
Book request	P	T	Refer to searcher	45D	Library user	Acquisition librarian	Preorder search	Order
Book request	P	T	Searched	45D	Library user	Acquisition assistant	Order	Data proc-essing
Book order (original copy)	P	T	Mailed	30D	Data processing clerk	Acquisition assistant	Order	Order
Book order (copy)	F	T	Filed in process	30D	Order clerk	Acquisition assistant	Order	Order
Book order (copy)	F	T	Mailed vendor	30D	Order clerk	Acquisition assistant	Order	Order
Book order (copy)	F	T	Filed public catalog	30D	Order clerk	Acquisition assistant	Catalog filing	Catalog filing

I = Informational
In = Instructional
P = Primary
F = Functional

T = Typed
W = Weekly
D = Daily
M = Monthly

FIGURE 5-4

SUMMARY WORKSHEET FOR THE SURVEY OF REQUIREMENTS									
System: Acquisitions—Accounting and Reporting Subsystem		Analyst:				Date:			
Name or Description of Requirement	Identify each output that is required to satisfy the requirement.	What operations prepare or maintain each of these outputs.			What activities receive each of these outputs both within or outside the system.				
Maintain and report distribution and expenditures of library funds.	Departmental control record	Accounting and reporting							
	School or account control record	Accounting and reporting							
	Monthly financial report	Accounting and reporting			Director	Assistant director	Dean of School		
Maintain and report current workload of orders.	Daily control sheets	Order subsystem		Serials system					
	Monthly statistical report	Order subsystem		Serials system	Director	Assistant director	Dean of School	Department representative	
Pay invoices	Cleared invoice	Accounting and reporting			Comptroller				
	Invoice control sheets	Accounting and reporting							

and reports maintained and prepared in support of each of the system's requirements. Another collection of output survey worksheets will detail the methods and procedures followed in preparing and maintaining the supporting reports and records, the sources of the information used in the outputs, and the records and reports used by designated individuals and operations to satisfy of the requirements of the system and of its subsystems.

A summarization for methodical study of the requirements of a system as well as of the records and reports purported to meet those requirements should be prerequisite to measuring the applicability of the data received by a system for meeting its requirements. The filled-out summary worksheet is organized by requirements and associated outputs within each subsystem. The example given in Figure 5-4 results from the analysis of worksheets of stated requirements and outputs of the personnel in the acquisitions system engaged in the major activity known as the "accounting and reporting subsystem." Parenthetically it should be understood that the example is but one of a continuum of summary sheets to be prepared by the analyst for the analysis of the acquisitions system.

The summary survey of requirements should supply the analyst with a general overview of a system, revealing the work flow, the interrelationships among subsystems, duplication of effort in preparing, maintaining, and supplying the system's outputs, and revealing uses and distribution of reports that may be unnecessary and can be questioned by the analyst in the light of the fairly complete picture of the system he has developed through the summary worksheets.

chapter six

FLOW CHARITNG

PURPOSE AND USE

Flow charting is the symbolism of the systems analyst. By laying out facts in a common sense way it enables the analyst to view graphically partial or complete systems and to interpret analytically what may have been recorded only as a narrative procedure. Through the use of standardized flow-chart symbols a middle ground is created between the librarian and the systems analyst and efficient and effective communication may occur as a result of a properly constructed flow chart.

Flow-chart symbols can be used to represent in a logical progression the elements (functions, decisions, and actions), the requirements, the inputs/outputs of the system, and the equipment used in the system. Therefore the flow chart illustrates the sequential flow of work and information through a system. The advantage of using flow charts stems from a system's being a process that is constantly changing; therefore in actual operation it is often difficult to understand without diagrams the relationships that exist within a system. The flow chart may be compared to a series of snapshots stopping the action within a system allowing the analyst, systematically and realistically, to evaluate the current operations and design new procedures if necessary. This is true because functions, decisions, and actions are clearly identified.

COMMONLY USED SYMBOLS

The IBM X20-8020 flow-charting template symbols have been employed in the examples. These symbols may be readily used in either a

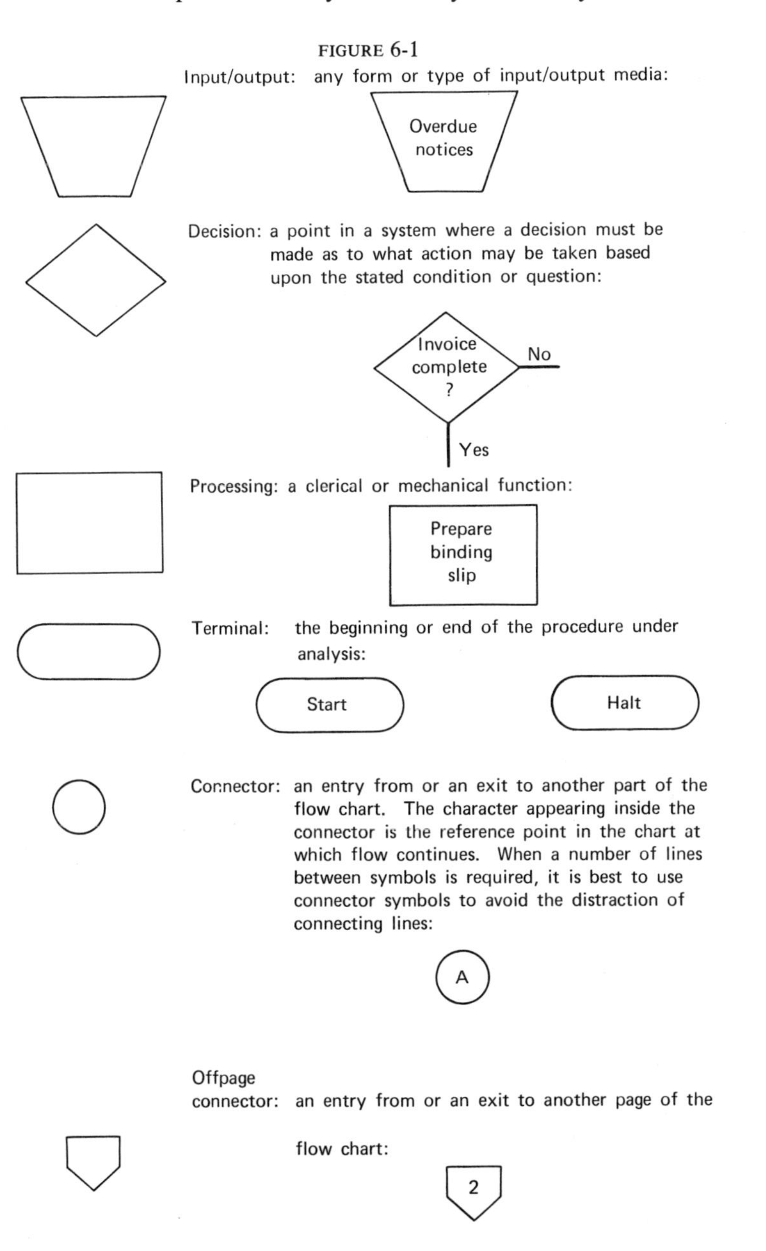

FIGURE 6-1

machine or manual system. The construction of the flow chart may be accomplished by one of two methods: (a) progression from the top to the bottom of the chart—vertical-flow method; or (b) moving from the left to the right in the chart—horizontal-flow method. Either approach has its merits and should not adversely affect the ability to interpret or construct the flow chart. However, once a method is adopted it should be consistently adhered to.

The six symbols are those most often used in flow charting (Figure 6-1.)

These common symbols used to show the sequential flow of work are connected by directional flow lines with or without arrows. Arrows only need be used when it is necessary to represent an exit or entry line from one symbol to another that is not in the same direction as the flow method being followed (Figures 6-2 and 6-3.)

FIGURE 6-2 Use of Connectors

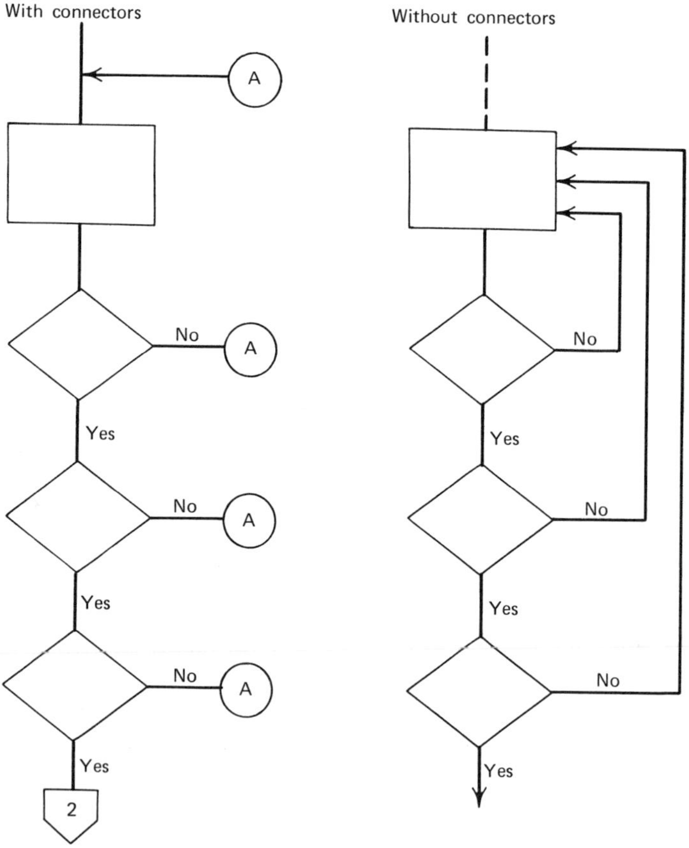

FIGURE 6-3 Use of Arrows

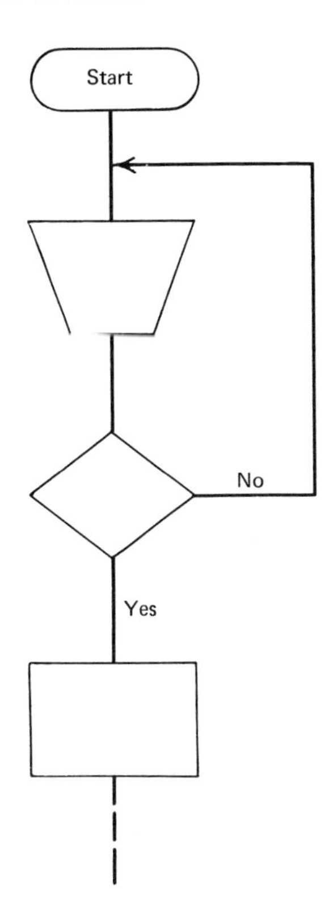

SPECIAL PURPOSE SYMBOLS

The following symbols may be used when more detail is desired or when the more specific symbols are readily applicable (Figures 6-4 and 6-5.) They may also be used in the flow charting of automated or computer systems.

FLOW CHARTING RULES

The following general rules should be adhered to when preparing a flow chart:

1. Conventional symbols should be used to facilitate mutual understanding of the logical flow of work.

FIGURE 6-4 Forms of inputs and outputs

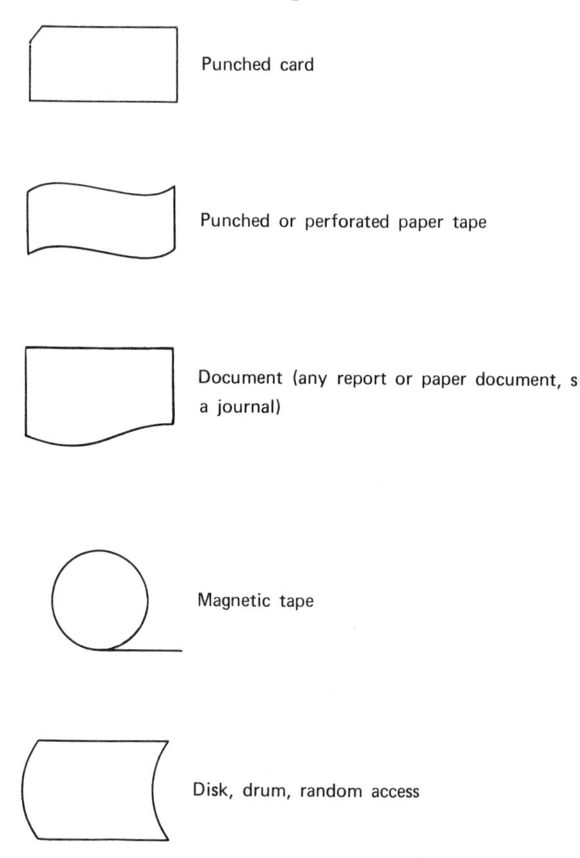

2. The system, or its components, or both, should have a clearly indicated starting and halting point in the chart.

3. The graphic flow of work should always be in one direction, normally top to bottom or left to right.

4. No directional flow lines should be unconnected at any point. Every directional line should lead to another step in the chart.

5. The descriptive statement within any symbol should be succinct and mutually understandable. The terminology used should be applicable to the system being studied.

6. Wherever ambiguity may be evident, annotation or side notes should be used to provide a thorough understanding of the various parts of the flow chart.

7. Each decision "diamond" should have two possibilities—a "yes" (positive) and a "no" (negative) path.

FIGURE 6-5 Processing functions

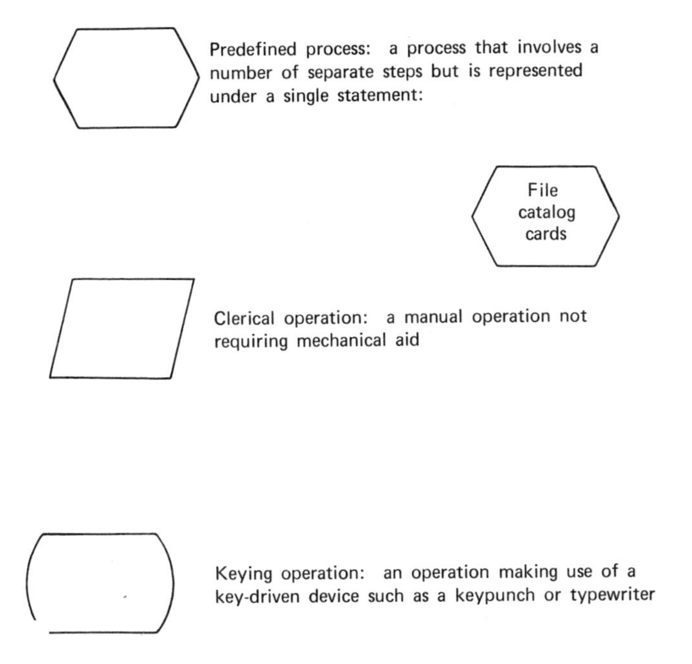

8. The flow of the work should be clearly indicative of what actually happens in the course of an operation. The realistic sequence of functions in the surveyed system should be followed throughout.

9. Because the flow charting of a complex system is an involved procedure, it is recommended that other staff members associated with the problem participate in the analysis phase of the flow-charting operation.

10. The analysis represented by the flow chart should be consistent within itself. That is to say a flow chart illustrating the progress of a document through a system should not be expanded into representation of a clerk's functions having no bearing on the operation being charted.

CONSTRUCTING CHART

The "make-ready" phase of a flow chart is of vital importance in creating a picture that is representative of the system. Awareness of the component parts of the system to be flow charted is obviously required. With the aid of summary analyses and an understanding of procedures flow charting is facilitated and the analyst should become aware of the distinctiveness of each process and the individual decisions and flow paths that exist under the present procedure.

In constructing the chart it is recommended that "pieces" of the system to be charted be written on individual slips of paper so that the initial arrangement of the flow-chart layout can be as flexible as possible. Once the arrangement is seen as satisfactory, the system can be transposed to sheet form. For expediting the often necessary photocopying of the finished flow chart a standard paper size such as 8½ x 11 should be used.

CAPABILITIES AND RESULTS

A flow chart pictographically representing a library system is uniquely capable of (a) assisting in the organization of the system's structure; (b) helping to visualize the system's component parts; (c) Resulting in improved techniques, especially since the finished flow chart can be utilized as an effective training device with significant advantages over written job descriptions alone; and (d) helping in communication about and understanding of the system.

At first in combination with the summary analyses and understanding of procedures the flow charts should show the existing system with all of its characteristic faults and strengths. One certain proof of the validity of an existing system is whether it can be flow charted effectively at all. If not, the system must be illogical.

By surveying the input to any entry point in the system the analyst as he proceeds step by step through the flow chart can see the uses that will be made of the input, what decisions must be made, how the input affects the decisions, and what action comes from each decision. Thus he can judge whether the decisions are correctly made in relation to the requirements placed against the system. He also can evaluate what effect various demands would have on decisions and their related actions in order to develop the set of criteria or managerial procedures that best satisfies the system's requirements.

With the standard rate for each function available, the analyst is in a position to analyze those operations appearing to be overburdened or near breakdown at certain steps depicted in the flow chart. He can then evaluate and suggest alternate paths to be taken, if any, to maximize the system's effectiveness. An evaluation representative of a "before" and "after" library binding procedure is shown in Figure 6-10.

The basic question to ask in evaluating each of the flow chart's parts would be the possibility of simplifying, eliminating, combining, or rearranging any of the elements of the work flow shown.

The results of the overall evaluation are to be used as the basis of a flow chart serving to suggest new and improved methods of accomplishing the operation under study. These suggested methods will involve major or minor changes according to the past operational success or failure of the present system. The new or improved methods will also depend on the resources available. If any automation is planned, a concept of vital importance in suggesting revision of a system is the establishment of manual procedures and operations that will be machine compatible.

FLOW CHART EXAMPLES

The remainder of this chapter is devoted to illustrative examples (Figures 6-6 through 6-11) of the flow charts of selected library operations to provide visual aid in further understanding how to apply the principles of flow charting. Figures 6-6 through 6-9 are progressively more complex while Figure 6-10 is a case study example of flow charting used to represent a library's current procedure and the resulting modified suggested procedure that after implementation proved to be a more efficient and effective operation. Figure 6-11, a generalized flow chart of a computer-based library operation, illustrates the use of computer system flow-chart symbols.

This chapter has been essentially a discussion of decision flow charting. Other charting and diagramming methods[1] may be used to describe paper work flow, functional organization, and work progress.

[1] For description of the various charting and diagramming methods and techniques see Systems and Procedures Association, Charting, in *Business Systems,* The Association, Cleveland, Ohio, 1966, Chap. 4; and R. M. Dougherty and F. J. Heinritz, The Flow Process Chart, Flow Diagram and Block Diagram, in *Scientific Management of Library Operations,* Scarecrow Press, New York, 1966, Chap. III.

FIGURE 6-6 Flow-chart figure—discharge of a library book

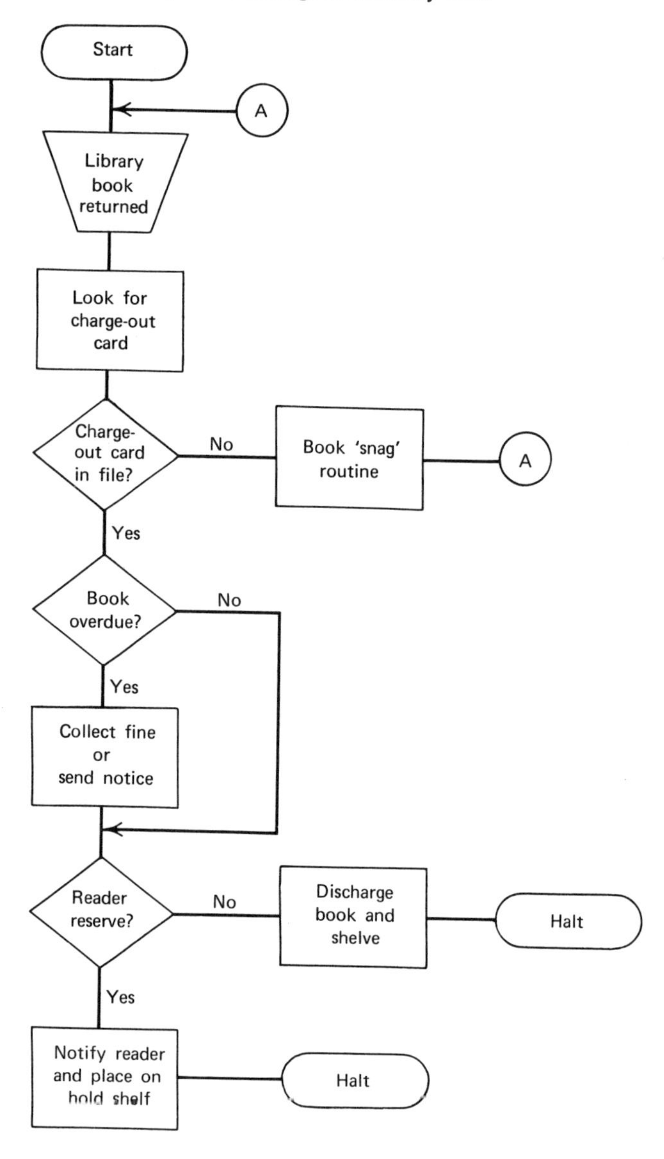

FIGURE 6-7 Flow-chart figure—clearance of an invoice

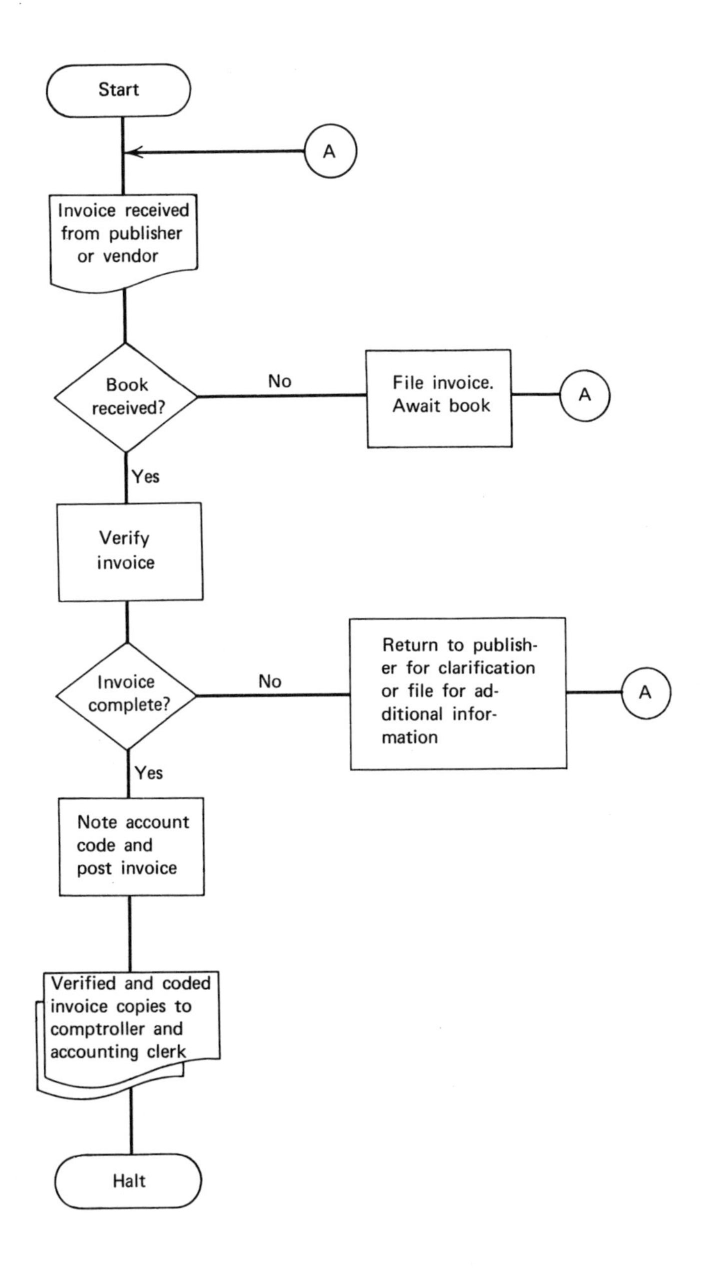

FIGURE 6-8 Flow-chart figure—check-in of a journal issue

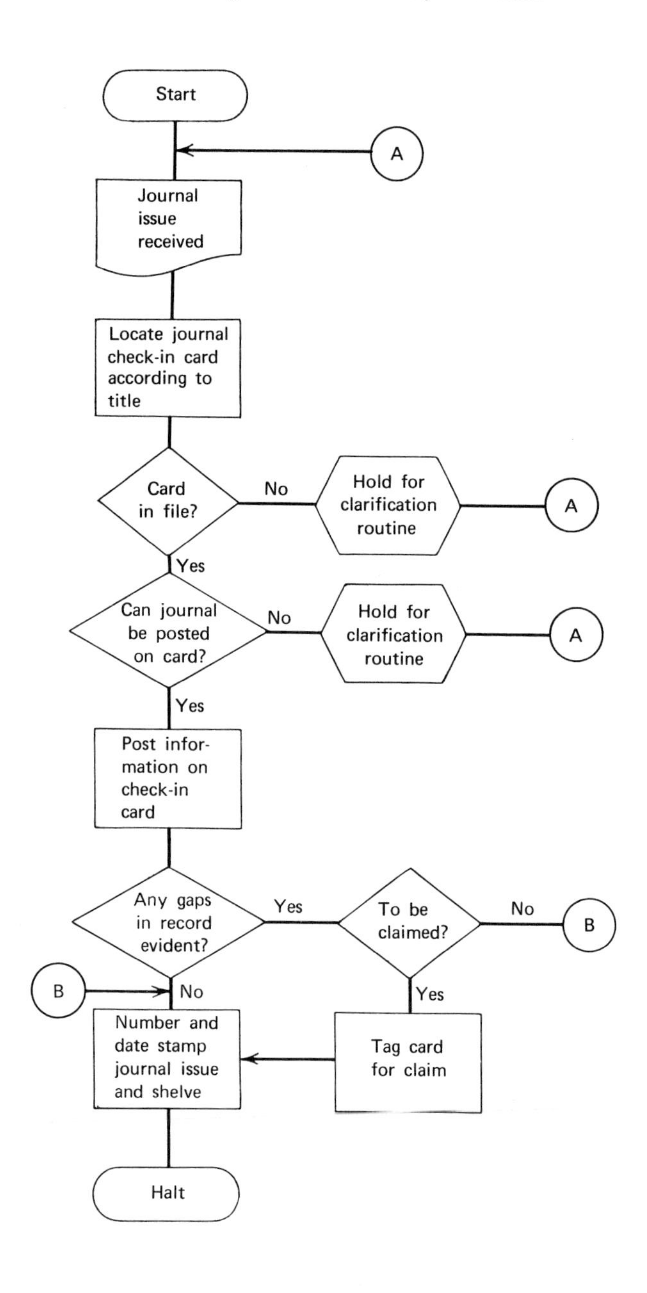

FIGURE 6-9 Flow-chart figure—processing of an added volume

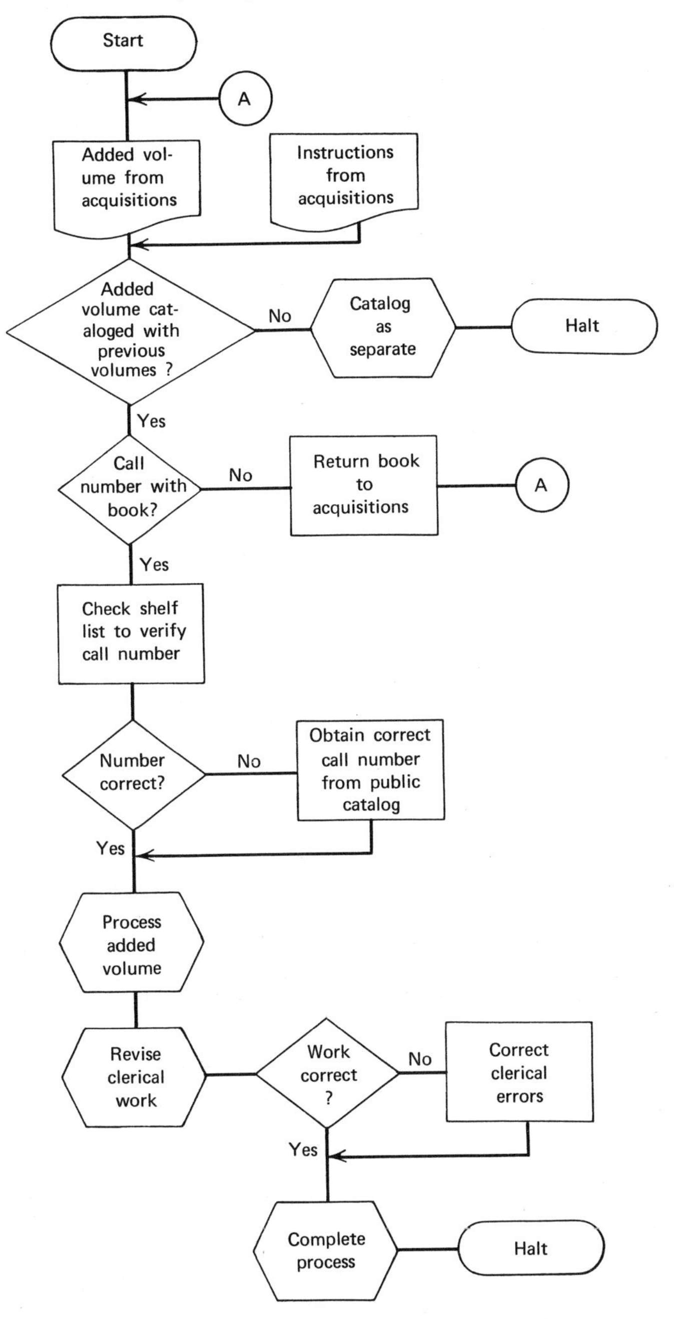

FIGURE 6-11 Flow-chart figure—photocopy service—automated invoicing and statistics keeping

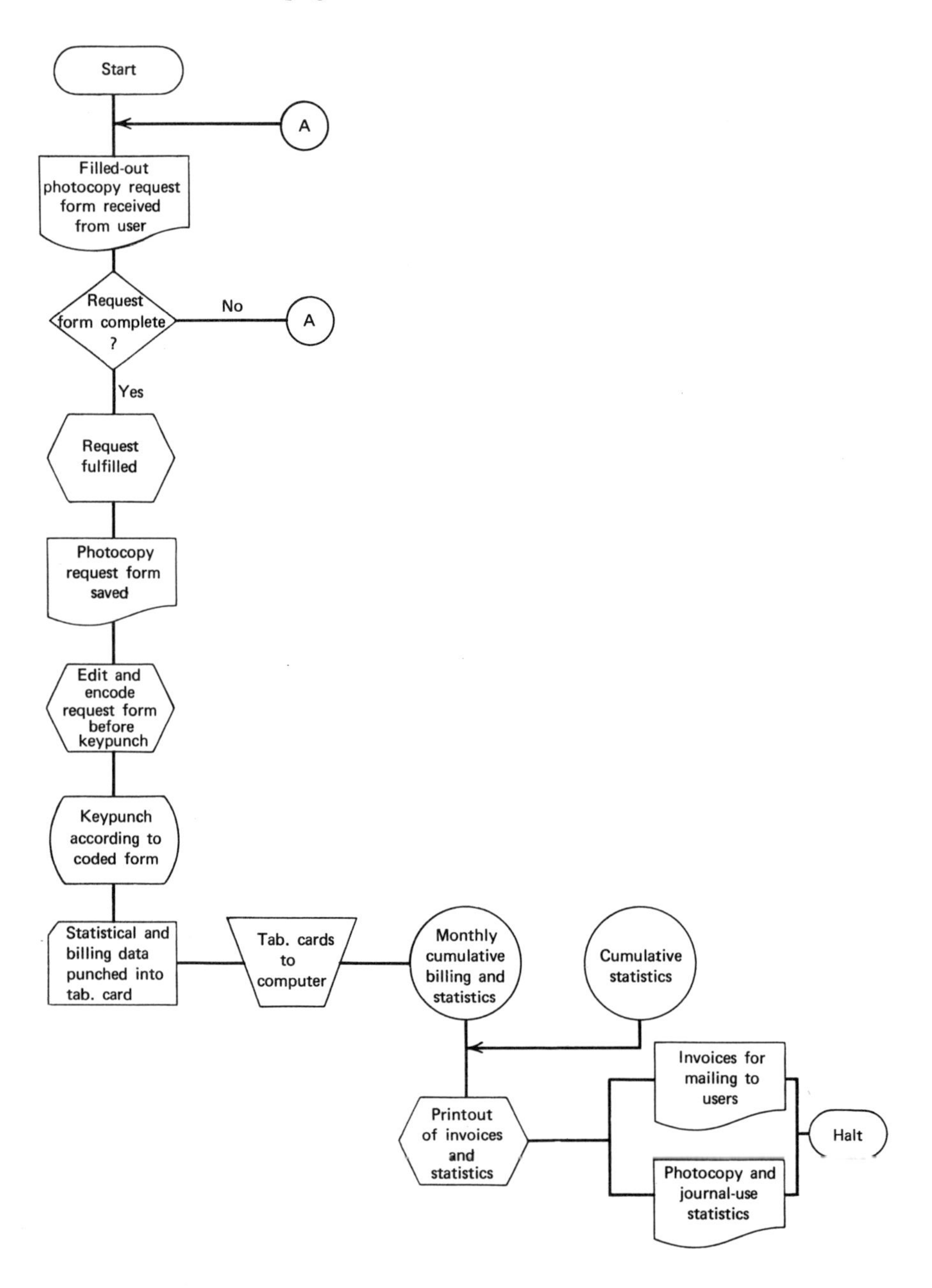

chapter seven

EVALUATION OF THE CURRENT OPERATING SYSTEM AND REPORT OF FINDINGS

APPLICATION OF MANAGEMENT CONCEPTS

With completion of the analysis phase the analyst is prepared to proceed with the second phase of the systems study—evaluation of how well the current operating system fulfills the demands placed on it. Based on a thorough understanding of present methods and procedures used in the current system, supported by the survey worksheets of requirements, personnel, equipment, the outputs and inputs of the system, together with completed sample copies of all forms used and records produced, the analyst is prepared to evaluate the system to determine how effectively, efficiently, timely, how accurately, and at what cost, the present system meets the requirements of each operation.

It is in this phase of the systems study that flow charting seriously comes into play.

Other techniques of scientific management applicable in library systems study are discussed in this chapter—the "management by exception" concept and the preparation of reports and recommendations in a persuasive manner and cogent form for consideration by management.

Statistical sampling and job analysis for developing work standards were briefly discussed in Chapter 4 in connection with the analysis and understanding of current procedures.

The evaluation phase culminates in a report of findings including recommendations for maintaining the status quo, eliminating or transferring certain operations to another system, reorganizing given operations, combining operations, increasing or decreasing staff, recasting the functions and responsibilities of professional personnel, or designing a completely new system.

The analyst should be aware of the options open to management on its receipt of the evaluation report. He should anticipate such contingencies as that management will give full acceptance to his recommendations, requiring the analyst to be prepared to proceed with the final phase of designing a new system or modifying the present one; that management may give initial disapproval of the recommendations, asking for further substantiation of them; or that management may place certain restrictions on the operations of the system, calling for review of the original recommendations for possible accommodation to such restrictions. These restrictions may arise from a need to economize, a change in management's goals, the elimination of one or more previously stated requirements of the system, or the abrogation of decision on the type of system (computer based or noncomputer based[1]) originally opted for by management.

In any event the adequacy of present procedures cannot be evaluated without thorough knowledge of what management expects the system to do, such as attaining maximum efficiency, highest productivity, maximum accuracy, and least operating cost. Because the confluence of these four factors, however acceptable and laudable as goals, is unlikely if not impossible, management is compelled to specify the primary goal. For example, if it is highest productivity management desires, it is likely that lower operating costs and maximum accuracy cannot be anticipated. Conversely, least cost is not entirely compatible with the goals of maximum efficiency, productivity, and maximum accuracy.

In the evaluation of current methods and procedures the analyst's report of findings and recommendations for revision, elimination, or maintenance of existing methods and procedures is based on specific determinants.

[1] The noncomputer-based system is defined as one employing conventional manual methods and/or any and all types of electromechanical equipment that may be activated to produce independently of a computer, required informational output by keyboard, punched card or paper tape devices.

1. There are requirements placed on the system that are germane to the successful operation and management of the system and of the total system of the library including those requirements logically contributing to the meeting of the requirements of related systems. To cite an example of the interrelationship of requirements among systems, a requirement of the preorder search subsystem of the acquisitions system is the supply of bibliographic information needed for ordering a title. In satisfying this requirement more frequently than not the searcher uses a source containing additional bibliographic data also needed in the cataloging system. The preorder search requirement should include the capturing of this data needed in the cataloging system in order to eliminate the costly, time-consuming repetition of the search.

2. There are inputs/outputs and controls of the system that adequately fulfill its requirements within the total system. Evaluation of the system's inputs/outputs is (a) to determine that information is received when needed for timely decision and action and (b) to eliminate the production and supply of information serving little or no purpose in meeting the demands placed on the system. The matter of balances in book fund accounts is taken as an example. Without accurate and timely reports of the status of accounts, control of expenditures in an orderly and equable manner throughout the budget year cannot be maintained. Similarly overexpenditure in given accounts can be discovered too late for any corrective action, leading to disruption of the system's functions and effective administration and planning. Obviously in the latter case the report management needs is one that indicates what and when individual accounts reach predetermined minimums for exercise of decisional control at the proper time. In further illustration book purchase requests submitted in variant form unrelated to the logical arrangement of bibliographic data best suited to searching and ordering procedures can only result in slowed and inefficient manipulation. What is needed here is submission of the book request in standard format.

3. Further determinants are those current methods and procedures judged adequate for processing the work loads. Again using the acquisitions system as an example management's goal is to attain a processing rate of 200 orders per week, which is at present not being achieved by current procedures and staff. The existing procedures must be evaluated with respect to their applicability and need in meeting the stated goal. On the one hand certain methods and procedures found unnecessary can be eliminated, thereby allowing the current staff to satisfy the desired processing rate; on the other hand, with present methods and procedures judged to be valid, additional personnel is indicated in order to reach management's objective.

4. The staff must be capable of fulfilling the system's requirements under the currently applied methods and procedures. Involved here are size of the staff and the capabilities, capacities, and skills or special training of individual workers. Depending on the results of staff evaluation additional workers with the necessary skills may be needed or perhaps workers should be reassigned to jobs fitting their levels of competency.

5. The equipment available must allow application of efficient methods. Changing, replacing, or adding a piece of equipment frequently leads to greater efficiency and improved productivity of a system. The application of efficient methods to required cataloging procedures by the addition of a simple piece of equipment is illustrated. Instead of manually copying cataloging data from the *National Union Catalogue* to produce needed cards for the library's catalog, a quick-printing enlarging camera can be procured to produce a near-standard image of the *National Union Catalogue* entry which in turn can be used as master copy to produce the number of copies of the card needed in cataloging by copying equipment. The addition of clerical help may be avoided by the increased productivity of the current staff through more efficient methods made possible by introducing one piece of relatively inexpensive equipment.

In arriving at the foregoing determinants the analyst uses the following "tools" obtained in the planning stage and analysis phase of the systems study: (a) managerial statement of goals enunciated in the planning stage of the study; (b) existing procedural manuals including job analyses and work samplings; (c) the worksheet prepared on each requirement; (d) preliminary survey worksheets; (e) worksheet prepared for each output; (f) summary worksheet prepared for all requirements and their associated outputs; (g) the worksheet prepared for each input; and (h) the summary worksheet summarizing all inputs.

When completely executed, the worksheets contain all the data the analyst needs for the evaluation of current procedures. It remains for him, prior to submitting his report of evaluation, to summarize systematically this mass of data supported by flow charts of existing operations.

Having come to some conclusion about the adequacy of the present system, the analyst should (a) compute the cost of processing a unit of material through the system; (b) measure the productivity of the system; (c) determine whether the system provides data promptly for timely action; and (d) evaluate the accuracy of information supplied.

During evaluation the analyst reviews the answers he has received in the analysis phase of the study and considers additional questions arising in his mind in the course of arriving at his initial conclusions. In testing his findings and conclusions the analyst may use the flow chart as a simulator of the system and its components. By following the flow of work through its associated functions, decisions, and actions he verifies or disproves the results that can be obtained under various operating conditions. This process, with examples, is now given.

1. Each component, as well as the overall system, should be appraised under increasingly stepped-up workloads to determine the maximum capacity of the system at which it can continue to fulfill its requirements. Does the present workload and accompanying cost meet management's goals? In the acquisitions system three factors patently affect the workload—the number of requests received, the number of orders processed, and the number of books received. The magnitude of workloads throughout the system depends on the volume of book purchase requests. By varying the number of requests entered into the system the ability of the subsystems (preorder searching, ordering, receiving, accounting, and reporting) can be evaluated and their maximum capacities estimated.

In the circulation system two factors affect the workload here—the number of books charged out and the number of books returned. The magnitude of the workload throughout the system depends essentially on the volume of books loaned. By increasing this number and noting its corresponding effect on the operations of the circulation system, such as the sending of overdues or shelving of books, the ability of the subsystems to meet the increased volume can be evaluated.

2. Current reporting requirements should be appraised to determine whether they stipulate statistical or other informational detail pertinent to the needs of management at various levels. Does management need all such reports or can the principle of management by exception be applied so that it receives only the information required for decision making or taking justified action? For example, in the cataloging system the supervising librarian need not be given a report on the size of the backlog of books to be cataloged so long as the size of the backlog stays beneath a predetermined level. As soon as the backlog has reached the problem zone management is notified.

3. Each input/output should be appraised for sufficiency in meeting the internal and external requirements placed on it. Do the outputs of a comptroller's office furnish sufficient financial data to satisfy the requirements of the acquisitions system? If not, what recommendations should be made to correct this? If a total system is one of management's goals, do the present outputs of the preorder search subsystem furnish the necessary bibliographic data required by the cataloging system and are the data furnished in ready-to-use form? If not, what modifications can be made in the form of the outputs of the preorder search subsystem; what effect would such modifications have on efficiency, capacity, and cost of the preorder search subsystem's operations; what effect would such

modifications have on the number and kind of staff needed in the subsystem?

MANAGEMENT BY EXCEPTION

Management by exception[2] is a technique of scientific management broadly applied in industry and business but generally not so consistently or consciously applied in the management of libraries. The technique is applicable at all levels of responsible action within the library's organization. It is the principle whereby management receives only that information on which action is indicated. To apply this principle it is necessary for management to be able to specify the information required for taking action and to set the point at which such action may be taken.

The timely supply of information about what *has not* happened under stated conditions rather than information on all that *has* happened necessitates a sifting out of the "exceptional" data from the mass of data not requiring action. Only that information signaling the need for corrective action, adjustment, and decision is supplied. Before management by exception can be applied, the conditions or parameters of control actually required for the exercise of assigned responsibilities and decision making should be precisely delineated.

The technique can find only limited application in manual procedures because of the inordinate consumption of time in manually reviewing data files item by item. However, the extent to which this technique can be used in computer-based systems is limited only by ability and imagination in defining the parameters of operational control needed. Among the following examples the objectives are not new but most of them suggest action not economically feasible by manual methods. The process of claiming issues of periodicals not received is an exercise in management by exception. Serials librarians have done this from time immemorial but the missing factor has been timeliness of action to assure reasonable receipt of the claimed issue due to laborious hand methods. If computer-based serials control is applied, a computer report may notify the serials librarian of all issues of foreign journals not received in a predetermined interval and produce the claim forms for these issues. As another example, the director may be troubled about a seeming overload in the acquisitions system. Trying to pinpoint the reason for an increasing backlog of book

[2] Aspects of management reporting by exception and management's informational needs are discussed in many sources including C. L. Littlefield and Frank Rachel, *Office and Administrative Management,* 2nd ed., Prentice-Hall, Englewood Cliffs, N. J., 1964, pp. 83-104.

orders, he does not ask for a record of all requests searched but just for those representing books already in the collection. Thus he and the acquisitions librarian by applying the principle of management by exception find that one third of the searches are for books already in hand or on order, indicating a significant reason for the overload in acquisitions. Again, the director of the library does not need to know the current balances of all book accounts each month but he does need to know in time for taking action that a given account has fallen to a predetermined minimum; or he may wish to see monthly reports on the balances of all book accounts beginning the seventh month of the fiscal year in order to decide whether the same rate of expenditures can be continued for the remainder of the year.

The concluding examples of the exercise of management by exception indicate computer rather than manual handling. The acquisitions librarian does not need a weekly report of all books ordered but rather a continuous report of items on order for a predetermined period and not yet received at the time expected. For keeping a measure of the extent of the cataloging backlog and for taking action on books for which Library of Congress catalog cards have not been received within a predetermined period, the catalog librarian needs a report of the number of books in process and a list of the titles to be originally cataloged in the absence of Library of Congress cards rather than a nonselective listing of all books in process.

In evaluating the system the analyst should test the reports received at each level of management to see if the principle of management by exception can be applied under current procedures and if applicable, what the effect is on the efficiency, capacity, and operating cost of the system. The design of a system should eliminate unnecessary reports and substitute reports of information actually needed for specified conditions requiring surveillance for action.

REPORT OF FINDINGS AND RECOMMENDATIONS

Finally, the study staff should submit to management a written report precisely stating its findings, conclusions, and recommendations. The study staff should provide enough detail to enable management to make the correct decision in relation to overall organizational goals. Depending on these recommendations and the decisions of management the study staff may or may not proceed to the culminating step of a systems study—that of design.

It is customary to think of reports as formal documents with such attending characteristics as formal language, exceedingly impersonal tone, great care in the making of claims and conclusions, inappropriate jargon, and inflated writing using redundancy and repetition. These characteristics can be obviated by using simple, direct language to state what needs to be stated, and no more.

Ideally the analyst in the actual writing of the report should have already, in a sense, written most of it because he has collected and organized all of the data needed to prepare the report of findings. He can follow his own investigation, can put together those parts of the report that contain information best known to him, and can finish the report by filling in any gaps from the information he has already recorded. The analyst should be able to state in specific terms in writing what the objectives of the study were and what was accomplished by the study. The accomplishments of any study should be statable in specific terms within the compass of a short paragraph of two or three sentences. This paragraph will become, then, the topic paragraph of the conclusion section of the report and as such is the most important statement in the entire report.

The task of filling in the gaps of secondary information will take a good deal of the time required for writing the draft of the report. This part of report writing can be difficult if the analyst does not have sufficient orientation to the organization of the report.

ORGANIZING THE REPORT OF FINDINGS

Some authorities seem to think it is detrimental to list the generalized outline of a report but if the writer will realize that the outline is generalized and that it should be used judiciously and changed when he believes it is necessary to change, a good outline is very useful. The outline follows:

- Front matter
    - Cover with title and identification
    - Title page
    - Tables of contents and figures
    - Foreword, preface, and acknowledgements
- Body
    - Introductory summary
    - Statement of problem
        - Objectives of study

Scope of study
Possible outcome
Method(s) of study
Planning and schedules
Personnel
Equipment and materials
Procedures
Results
Collected data
Interpretation of data (reduction)
Method(s) of data analysis
Conclusion
Conclusions
Recommendations
Implications
Back matter
Appendixes
Detailed data sheets
Computations
Large figures and drawings
Appended documents and correspondence
Bibliographic matter
Index (only if report is voluminous)
Back cover

SEQUENCE OF WRITING

One of the hazards resulting from an outline such as the one above is the inference many people draw that the actual writing of the report should be done in the order indicated. A more appropriate order is suggested here.

The first piece to be written could be that labeled "Statement of Problem," and in particular the subsection "Objectives." The next section should then be Conclusions, followed by Recommendations. Then the Results section can be sketched out fairly fully and Method(s) come next. From here on the order of the writing does not make as much difference as in the earlier parts, with two exceptions. The Introductory summary should be very nearly the last section to be actually written. This section, frequently called "Abstract" or simply "Summary," is the most-read part of the report and should therefore be the best. The last thing, usually, is putting together the table of contents and other similar tables. The table of contents should be started with an outline such as the one here, with appropriate changes made to that outline as the report is developed, and with the notion that the outline will indeed become the table of contents by the addition of page numbers.

POINT OF VIEW

The report writer needs to know what point of view he should use in the writing. It is most likely that he will need a dualistic point of view: first, he is the expert communicating to peers; second, his writing should be organized from general to particular; it should have the inverted pyramid style of the news reporter. The writer of the report has done the investigation that he is reporting; he should know more about it than anyone else. On this basis alone he is the expert. The general-to-particular organization of the writing itself is called for because it is true that more than one person will make real use of the report and that these uses themselves will vary and that any one reader will not have to or want to read all the various sections of the report. If every section and subsection is so organized, all readers will be able to grasp the total import of the report without having to struggle through every paragraph.

The best device to indicate the orientation of the report to the reader is the liberal use of headings. The organization and selection of these begin with the early outlining of the report. Just as the outline for the report becomes the table of contents, it also serves as the basis of the list of headings. The writer should take advantage of this in checking several times during the actual writing to see that he is using the outline to indicate headings and to check and change his outline as he includes additional headings. An excellent discussion of the formatting of headings is included in Ulman and Gould.[4]

FUNCTIONS OF PARTS OF THE REPORT

Following are brief statements of functions of the individual parts. The order here is that of the outline listed earlier.

Cover. Two functions are performed by the cover. It physically holds the report together and provides sufficient identification for the

[3] J. N. Ulman, Jr., and J. R. Gould. *Technical Reporting,* Rev. ed., Holt, New York, 1959, pp. 17-19.

locating and handling of the report as a document; the cover should provide for easy access into the report itself through convenient binding in terms of the thickness of the report.

Title Page. The title page contains the informative title and complete identification as to author's name, position, organization; date and place of publication; place of the report in any series, including serial enumeration, if any. It is usually desirable that all the information on the title page be included on the cover.

Table of Contents, Illustrations, Figures. Informative, complete, and highly utilitarian guides to the reader as to location, parts, size, and interrelationships are supplied by the front material tables. Such tables should be made as legible as possible and should occupy as much space as is needed to assure completeness and legibility. They should not be crowded.

Foreword, Preface, Acknowledgements. These sections are considered optional; they should be used to indicate something about the justification for the writing of the report, any special circumstances that tend to make the report different from what the reader might expect, and any unusual help given the writer/analyst. If these sections are used they should be kept brief.

Introductory Summary. Most readers appreciate a summary that is not a linear abstract. The summary should give short versions of the important parts of the report, particularly the objectives of the study, the method of study (this should be very brief), the specific results, the conclusions, and the recommendations. If the writer has followed carefully the general-to-specific order of writing, the beginning sentence or two from each of these sections will serve as the essence of the summary. Many writers indicate that the summary can best be written by simply repeating such sentences verbatim and providing adequate transitions. The introductory summary should be considered by the writer as a document that may have to stand in stead of the entire report; it is therefore a very important part of the report and should be worked on with great care and a good deal of energy and thought.

Statement of Problem. This section defines the study (it should not define the report except as an incidental matter). It is probably the second most important part of the report.

Method(s) of Study. In many reports this is an invaluable section but if the method is routine and well known, the section may be quite brief and even nonexistent.

Results. Although many report writers consider this section as very important, often it is not and frequently it is not even read thoroughly, if at all. One of the problems here is that "Results" is confused with

"Conclusions." The only practical function of this section is to state in abbreviated form the actual data collected, so that the reader, if and as he wants, may verify the conclusions from the actual observations.

Conclusions. The conclusion functions as the answer to the question, "Precisely what was accomplished by this study?" If this question is not well answered, the reader is likely to gain the impression that not much, if anything, was accomplished by the study.

Back Matter. Customarily the back of the report serves as the repository for all the other material that did not seem to fit into the body of the report. This is too often the case. Instead, the back of the report should contain items of information that may be of good use to some readers. The analyst should attempt to make a reasonable judgment as to the potential utility of the material and if in doubt should not include it.

USE OF ILLUSTRATIONS AND GRAPHICS

If the information to be presented concerns primarily a description of a space, such as a reading room or a set of offices, and if the information is primarily dimensional, a dimensioned drawing such as a floor plan is more communicative than words. The argument can be carried on to other characteristics besides dimensions the conclusion being that other forms of illustrations (sketches, perspective drawings, flow charts,) and so forth may prove superior in their communicability to words. When this is the case, the "nonverbal" may be used first and supported by verbal interpretations.

A table containing specific numerical values is easier to use than a graph from which certain values must be derived. If the specific numerical values are not the major concern but trends and comparisons are, bar charts, column charts, and even pie charts are much more communicative than tabular material.

Most illustrations and graphics should be introduced with text material and it is probably best that the writer follow the nonverbal with a verbal interpretation. This means that the nonverbal material will be worked into the text of the report linearly—this is, so that the reader is almost forced to study the nonverbal as he reads the verbal material. Illustrations and graphics should not be located in other parts of the report, even if only a page away. A discontinuity occurs when this happens and when the reader gets to the page where the nonverbal material is, he tends to look immediately at it rather than at the continuation of the verbal material he is reading.

Finally, nonverbal material should be presented in good style. The art work should be neat and professional. Legends should be plentiful, accurate, and informative. In regard to the use of mathematical expressions over English, Kapp states ". . . every writer, before embarking on mathematical forms of presentation, should ask himself the question: is your mathematics really necessary?"[4] Keeping in mind who the readers of the report may be, the use of mathematical formulas may not be the best means of communication. Rather their use may complicate decision making by management to implement the report's recommendations.

Douglass,[5] Schmid,[6] Sigband,[7] Tichy,[8] Ulman[9] and Wyld[10] are considered excellent treatments of report writing and contain good bibliographic material.

[4] R. O. Kapp, The First Draft, in *Computer Peripherals and Typesetting,* by Arthur H. Phillips, HMSO, London, 1968, App. 3, p. 611.
[5] Paul Douglass, *Communication Through Reports,* Prentice-Hall, Englewood Cliffs, N. J., 1957.
[6] Calvin F. Schmid, *Handbook of Graphic Presentation,* Ronald Press, New York, 1954.
[7] Norman B. Sigband, *Effective Report Writing for Business, Industry, and Government,* Harper and Row, New York, 1960.
[8] Henrietta J. Tichy, *Effective Writing for Engineers, Managers, Scientists,* John Wiley, New York, 1966.
[9] J. N. Ulman, Jr., and J. R. Gould, *op. cit.*
[10] Lionel D. Wyld, *Preparing Effective Reports,* Odyssey Press, New York, 1967.

chapter eight

PRINCIPLES OF SYSTEMS DESIGN

In designing a new system the same data used in preparing the report of evaluation of the existing system are used in the design phase that culminates the systems study. The evaluation report obviously serves as the starting point or basis for the design of a new system, containing as it does suggestions for substantive changes by the analyst who at this point in the study should be conversant with the strengths, weaknesses, and unsolved problems in the present system and should be fairly well convinced about the corrective actions to be taken.

REVIEW OF GOALS

The goals and objectives of the library's management should be reviewed in relation to the requirements currently being placed on the library as well as in relation to the library's long-range plans of development and those of its governing administration. These goals, after being reaffirmed, amended, or changed, can be accepted as definite and the requirements of the existing system may be re-evaluated against them. Are extant requirements still pertinent? Can specific requirements be changed if this is needed? Answers to such questions cannot be arrived at arbitrarily but should be developed in the light of pro and con views of the responsible staff members who are to make the new system operable. Any desired changes in requirements controlled by outside sources would be subject to consultation with such sources.

ECONOMIC FEASIBILITY

In addition to confirmation of the goals of the library the analyst should secure affirmation of the economic parameters that must be taken into account in designing the new system. If in his preliminary decisions about what corrective measures should be taken the analyst sees the feasibility of computer applications, will the library and its governing administration approve the additional cost of electronic data processing including the cost of peripheral equipment needed for library computer-based operations? If additional personnel is suggested for increased productivity and elimination of chronic backlogs of work, is the administration prepared to meet the increased operating costs? These and similar factors of increased costs should be delineated before attempting the design of the system if it is to result in one fitted to stated cost limitations rather than one providing an apparently ideal solution.

In evaluating a proposed new system management's decision to accept it is strongly influenced by cost of installation and operation; that is, staffing, equipment, space, new forms, and other expense items. Consequently the proposal to management must include a comparison of the total and, more important, the unit operating costs of the current system with the projected costs of the proposed new system; it must also include the conversion expense and any capital expenses entailed in remodeling and expanding working space.

The cost of changing over from a current system to a new or modified one are always significant and often are not sufficiently analyzed and evaluated for valid managerial decision. Estimated time (costs) and the cost of units of new equipment to be blended with existing equipment required for conversion are necessary in any case but entirely imperative in the case of converting to computer-based operations. Underestimation of the time needed for records' conversion and the auxiliary equipment and staff required may lead to considerable embarrassment and even to failure in implementing the proposed system.

As an illustration of the seriousness and impact of the time-cost factor, let us assume that the analyst recommends and management concurs in the adoption of a computer-based cataloging system for the library. Such a decision can be ill founded unless supported by an estimate of the comparative cost of continuing the cataloging by traditional methods and the cataloging by computer-assisted methods; and it should be further supported not only by an estimate of the cost of converting the present catalog records for a beginning base book catalog but also by a projected estimate of the annual costs to be incurred in keeping the machine readable record up to date and the book catalogs in print and cumulated.

In recommending any computer-based system it should be realized that the conversion of man-readable records to machine-readable records may be the most expensive aspect of the program. The next most expensive factor generally is the printing and publishing of the machine output. It is, incidentally, these two factors that call for application of the principle that machine records should contain only data clearly essential to meeting the library system's requirements.

To continue with the illustration the analyst can readily make a close estimate of the annual cost of personnel, supplies, and other expense items incurred in the existing cataloging operations. The proposed computer-based system contains the same cost elements but with the addition of programming; personnel, who may be staff members available because of discontinued manual functions, for preparing the computer input; purchase or rental of inhouse input machinery; subscription to the Library of Congress MARC magnetic tapes; computer processing time; and the printing and binding of the cumulating catalog records during the year. Programming costs are usually very difficult to estimate, mainly due to two factors: (a) the individual programmer's ability and knowledge of library requirements and (b) the complexity of catalog filing and catalog data. Computer processing costs will vary with the program's execution time in producing a required listing. To a considerable degree the computer time costs will be affected by the preciseness of the program's logic and the size of the data base. In estimating the printing and binding costs dependence may be placed on a printing house representative.

It can be anticipated that annual cataloging personnel charges will remain approximately the same. Supplies and expense items including new equipment rental or amortization, except for on-line input devices, will tend to be less than in a manual system. The estimated additional annual operating costs are offset by savings such as lesser cataloging costs up to the computer-connected operations; elimination of the need to purchase additional catalog cases as the collection grows; the freeing of centrally valuable floor space taken up by the catalog cases; elimination of the preparation of duplicate sets of catalog cards for departmental and subject-area catalogs; the possible sale of printouts from the MARC tapes to regional institutions as aids in cataloging, current book selection, and collection evaluation; the processing of books more speedily to the shelves; the saving of the time of library users and the staff with the printed book catalogs being available at various points within and outside the library; and the possible sale of the printed book catalogs to other libraries.

Referring again to estimation of the extraordinary annual expense of computer processing and press printing of book catalogs during the year, the yearly pattern of printout and publication consistent with adequate service to readers must be determined—monthly, quarterly, four-month, or semiannual cumulations. The monthly cumulation pattern obviously is the most costly although it does provide near-maximum user utility outside of on-line accessibility to the catalog data in the computer data base. The cost of the monthly pattern will be found to be in the order of twice as great as that for any other of the alternates.[1] It further will be found that the variation in cost among the quarterly, four-month, and semiannual cumulations is not significant. Although less convenient to the user than the monthly cumulation, the quarterly pattern including two noncumulated monthly catalogs in each quarter results in the reader having to consult from two to a maximum of five printed catalog listings until the annual cumulation is supplied.

The costs of converting and printing the existing catalog records as the basic book catalog in the "startup" of the computer-based system represent an extraordinary one-time expense with no offset savings over manual cataloging methods because the card catalog must be maintained until the base catalog is printed and published. The costs to be estimated here include the cost of preparing and "loading" the catalog data into the computer, the cost of computer processing, and the cost of press printing the base catalog in the edition size found to give the most effective service to users. Preparation costs of catalog records for the computer include estimation of the hourly rates of the records' input keying operation, proofreading, and correcting, as well as an estimate of the time required for inputting the records into the computer.

To complete the analysis of the cost involved in his recommendation to install the computer-based cataloging system, the analyst using the cost data gathered to this point should prepare a five or six-year calendar of computer and publishing costs including the one-time base catalog cost in the first year of the operation of the computer-based system. A pattern of catalog cumulation must be proposed: single-year cumulations

[1] *Feasibility of Computer-Produced Book Catalog vs. Manually-Produced Card Catalog,* Unpublished typescript study, Rensselaer Polytechnic Institute, General Library, June 25, 1968.

to the fifth year when a five-year cumulation would be produced; annual cumulations on a year-to-year basis; or other patterns judged appropriate to the economic capabilities of the library and that will give adequate service to users. Given an anticipated annual average acquisitions rate and using currently prevailing personnel and machine costs, the analyst can extend this analysis over a period of years showing that the annual costs of records' conversion, computer processing, and press print will remain more or less constant. It, however, can be anticipated in a computer-based cataloging system, that net annual costs will be greater by approximately 25 percent than those incurred in following conventional manual procedures, excluding any of the possible offsets mentioned earlier.[2]

An analysis such as that descibed above should be undertaken for all proposed computer-based systems if the governing administration is to reach a valid decision on whether or not to adopt computer-based methods for the library. Computer-based or noncomputer-based systems design should be validated by analysis of the time-cost factors involved.

A major factor in management's decision to accept a newly designed system will be the improved ability to satisfy the system's requirements. This improvement may be beneficial to a related system within the library, to a coordination of systems, to library users, and to management decision making. Although these benefits are usually too intangible to be measured in dollars and cents, they should be submitted in justification of costs. For example, improved service to library users, improved quality, timeliness and form of information and reports required for decision and action, and improved coordination of operations among systems may be cited to justify a computer-based circulation system.

UNIT COSTS

The cost of the processing of each unit of material in each system is necessary in a comparison of the total operating cost of the former systems with that of a newly designed one. Comparative total costs do not supply management with a basis for making a decision to accept the proposed system. Determination of the cost of processing one book order, of cataloging one title, of discharging one book, or of processing a similar work unit of a system, correlated with the standard rate of processing each unit, should yield meaningful total costs' com-

[2] *Feasibility of Computer-Produced Book Catalog vs. Manually-Produced Card Catalog, loc. cit.*

parison in relation to the rated capacity of the old system and the required capacity of the new. Therefore, management can decide to accept increased total operating costs, aware that a stated requirement for increased productivity legitimately causes the differential in operating costs.

This discussion on unit cost of processing is confined to the major cost factor of directly chargeable personnel within a given system. While this, in general, is sufficient for establishing unit costs, it may be refined and in certain cases should be: for example, the shared use of workers between systems with each chargeable at a prorated cost and the shared use of equipment whose rental and operating costs also are proportionately chargeable.

ELEMENTS OF DESIGN PHASE

The design phase of the systems study includes nearly all the elements of the analysis phase, varying only in objective and approach. It will be recalled that in the analysis phase a beginning was made by determining the procedures and methods currently employed by the system, followed by analyses of the requirements, outputs, and inputs found in the system. In the design phase the beginning process is validation or determination of the requirements to meet management's goals, validation or revision of the system's outputs, and validation or change of controllable inputs. Based on the results of these decisions new operating procedures are promulgated to fit the needs of the system in meeting its requirements. Thus the order of approach in systems design is verification of requirements and then specification of the systems' inputs/outputs. Greenwood[3] supplies a thorough outline of the analyst's tasks in designing systems.

There is also a strong element of parallelism between the design and the evaluation phases. In evaluating current operating procedures and methods the analyst is observing and seeking evidence of the efficacy as well as the inadequacy of applied procedures and methods. In the process and as a logical by-product he is gradually formulating and substantiating his ideas for needed change, revision, or scrapping of specified procedures, anticipating the design of a new system. Directed to design a new system the analyst retraces his steps through the system in the evaluation phase. He checks his findings and prepares specific recommendations for

[3] James W. Greenwood, Jr., EDP: *The Feasibility Study-Analysis and Improvement of Data Processing,* (Systems Education Monograph No. 4), Systems and Procedures Association, Washington, D.C., 1962, pp. 22-27.

recasting procedures for organizational structure and job descriptions, for redesigning forms and reports, and for improving working-space conditions.

OBJECTIVES OF NEW DESIGN

Among the several objectives in designing a system perhaps the major one is planning for ali functions of a system to logically fall within the purview of the system. Other important objectives are that there be correlation and interdependence of all functions, subsystem to subsystem and worker to worker; that similar correlation exist in the preparation and maintenance of records and files and that unnecessary duplication of records and files be eliminated; and that the systems' activities be performed in a logically sequential mode, organized physically for the smooth flow of work through the system.

Another significant objective in systems design is definition of the authority needed by the head of the system as well as by the supervisors under him in order that all staff members may make the decisions and take the actions indigenous to the operating effectiveness of the system. Corollary to this objective the analyst should designate the one supervisor to whom certain workers in a system are to report. The establishment of this relationship is necessary to prevent worker disorientation and the consequent hampering of the orderly flow of work designed into the system.

Having a knowledge of the primary requirements of the system the analyst-designer should specify what ancillary requirements, if any, are to be met and the outputs necessary to satisfy all the requirements placed on the system. For each new or modified requirement and output a new or corrected worksheet for each requirement and output should be filled out as fully as possible. Although the same worksheet forms are used, the difference lies in the analyst's approach. Here he is specifying the requirements to be met and the outputs to be prepared in meeting those requirements instead of simply recording the *status quo* of requirements and outputs called for in the initial analysis of current methods and procedures.

If the new design is restricted to available capabilities in the present staff, procedures should be designed to fit these capacities and levels of ability. It is particularly important that procedures be as routine as possible in order to limit the disruptive effect of any exceptions in the preparation of data and records standard to the system's operations. The same is true if the new design is to make use only of available equipment. If restricted to the use of typewriters by budgetary exigencies, for example, there is no point in designing a system to exploit the characteristics of a paper-tape typewriter or key punch.

MANUALS OF PROCEDURES AND OPERATIONS

A cardinal principle in designing a system is that all procedures developed should be put in writing. As procedures for each operation are developed they should be described in sequential detail in writing, tested and supported by flow charts graphically illustrating the procedural steps in each operation. These detailed descriptions of the procedures to be followed in the various operations are, of course, the basis of integrated manuals of procedure covering the system's major activities and fixing the logical flow of work through the system. Appendix I to this chapter contains an illustration of written procedures or operations and a flow chart covering the preorder search subsystem in the acquisitions system. Appendix II contains an illustration of an integrated manual of operations that is designed for the total library system. It is constructed to reflect the administrative structure of the organization, the procedures within operations, and the forms used with a record retention policy statement for each form. The program, policy, and procedure for processing the library's photocopying office charge accounts are outlined in detail.

MANUAL PROCEDURES

The design of appropriate manual procedures is a principal prerequisite in the design of any type of system whether it involves electronic data processing or not. In any case and even in the most competent of computer-based systems there is substantial use of manual procedures in gathering information, in preparing this information as input to the computer system, and, as necessary, in processing the outputs to satisfy the requirements to be met by the system. In designing the system, however, it is essential to know whether methods and procedures will be based on the use of common business machine aids or on the use of the computer and peripheral machine aids.

The analyst should be thoroughly familiar with the various types of business machines that can be used in support of sound manual procedures in a system employing punched card, paper-tape, accounting, or copying equipment. At the same time he should be conversant with the techniques of computer-based systems as well as the supporting manual equipment compatible with computer input/output operations.

Manual procedures systematically and logically developed for achieving maximum efficiency in a noncomputer-based system should supply a satisfactory basis for building a computer-based system should this become feasible in the future. At that time programmers will need to reanalyze in minute detail the procedures considered adaptable to computer manipulation for instructing the computer in the step-by-step accomplishment of tasks in substitution for manual processes.

CRITERIA OF PROCEDURES

Good procedures possess certain characteristics that should be kept in mind in the course of designing a system. As the analyst develops procedures for implementing a system he should raise the following questions about each:

1. Does it furnish the outputs needed in satisfaction of stated requirements of the system?
2. Does each step in the procedure have a definite purpose in the generation of the output? Is it clearly necessary?
3. Does each function in the procedure move the item being processed one step nearer its completion as an output of the system?
4. Are operations staffed to achieve balance among procedural steps resulting in a smooth work flow and prevention of bottlenecks?
5. Are nonroutine processing steps reduced to a minimum with a path minimizing disruption and delay provided for any exceptions?
6. Are available mechanical aids used to the extent possible in satisfying the objectives of the procedure?
7. Are the forms used in the procedure so designed as to supply data in a logical format common to related functions and systems, saving time of execution and obviating annotation in their flow through the system?

Satisfaction of the criteria implied in these questions is fundamental to the successful flow charting of the procedures.

DESIGN OF PRINTED FORMS

The analyst in the course of the inputs/outputs survey has gathered samples of all the forms used and by the time he begins the design of the new system he will have decided on the forms or copies of forms to be eliminated, changed in format, and created. Because forms are the means of transmitting information and storing and avoiding repetitive transcription of information, the importance of the proper design in data processing is obvious. It is equally important that the form be designed for convenience and time saving in its preparation and use. The impact of the time lost in working with ill-designed forms is also obvious.

The design and redesign of forms is influenced and guided by several requirements and decisions.

1. Does the form provide for the recording of information in an itemized order consistent with the logical ordering of the elements of information recorded, permitting the most convenient preparation and use of the information at the least cost?

2. Does the form best serve the reciprocal and correlative characteristics of recorded data? In illustration of both Items 1 and 2 the book request form should be designed for recording bibliographic data during the preorder search in itemized progression standard to bibliographic entry; it should serve both for the purposes of ordering a book and for being a record for the originator of the request—thus indicating a two-part form. The book ordering form should provide for itemized arrangement of data in the exact format of the book request and should be in multiparts to serve as an order slip, vendor's packing slip, in-process slip, request slip for Library of Congress cards, the on-order record in the public catalog, and the notice slip to the originator of the request notifying him that the order has been received.

3. How will the form be prepared? If it is to be handwritten, item spacing must be sufficiently generous to avoid unintelligible abbreviations and crowding of characters. If it is to be machine prepared, machine spacing and consistency of item starting points permitting maximum use of the tabular setting should be designed into the form. If both handwriting and machine preparation are involved, the form's design should be flexible enough for both methods. A principle here is the minimizing of handwritten data, confining it to the checking of boxes next to specified information printed on the form. The use of check boxes wherever possible not only speeds preparation and use of the form but may also improve the accuracy of the record due to the simplicity of checking a specific box.

4. Will the form be the basis for a temporary or permanent data file? Standardized sizes should be adopted fitting the filing equipment usually available in a library. If a sheet form is required, it should be designed with adequate binding margins, punched if for ring binders, and specified in standard size for economy in printing and photocopying.

5. How many copies of the form are really needed to transmit information? For example, a requirement to be satisfied by the cataloging system is notifying the requestor of a book that it has been received and is ready for his use. The copy of the book ordering form used for ordering Library of Congress cards is returned and is used to notify the requestor reducing the multipart form by one.

6. Should commercially available stock forms be adopted? The adoption of such a form more frequently than not necessitates compromises when applying the form to local informational needs. Apparent savings in the purchase of stock forms can result in increased costs of processing. Forms designed to meet the data needs for specific inputs/outputs of a given system are more economical, useful, and efficient than standardized forms designed for routine and purportedly common applications. Forms should not be regarded as inviolable but should be constantly re-evaluated with respect to their usefulness and ease of preparation. Whenever justified, they should be redesigned and replaced for more economical processing that will more than offset the cost of new ones and the discard of the inventory of the old.

7. What are the criteria in designing the multipart form? The first factors to be considered are the reciprocal and correlative characteristics of the data to be carried by the form—that is, the number of copies to be transmitted to other points. If such a multipart form is indicated, it should be printed in a continuous perforated strip for economy in processing achieved by elimination of repeated machine insertions and alignments. The advantages and disadvantages of carbon inserts and self-carbon papers should be studied, as well as the stubbing of the continuous form by adhesive, stapling, and other means for holding the alignment of copies. For example, in machine operations stapling can prove injurious to the equipment as well as cause misalignment of copies of the form. The arrangement of the multipart form should allow sequential extraction of copies in the order of the uses to be made of them. Further, the arrangement of copies should provide for the distribution of the original and the clearest carbon copies to the points where the data are used most. If instructions relating to certain parts of the form are needed, these spaces should be referenced to the instructions by number or other coding. The use of different colored copies of the multipart form should be investigated for identifying the origin of the form. If a copy of a form is to be mailed, it should be designed for insertion in a window envelope.

8. Is the form clearly understandable? The form in all its copies should be identified by headings clearly indicating the particular purpose and use of each copy. The exact data to be entered in the form should be self-explanatory. For purposes of routine distribution the various

copies should be identified by numbers, bold-printed titles, or different colors of paper stock. As required the form should carry on its face or verso printed instructions for its preparation or use.

To summarize the foregoing considerations and decisions in the design or redesign of forms, the following elements should be sought: clarity of layout requiring a minimum of printed instruction; handwriting held to a minimum with checking of printed captions maximized to reduce errors and save time in entering information; arrangement of items in a sequence parallel with the functions to be performed; appropriate spacing for convenient entry of the required information; and a size for easy handling and for economy in printing, preparing, photocopying, filing, and mailing.

The design or redesign of forms is an inescapable task in the design of a new system. The function of the analyst is to apply the principle that the same data or information should be recorded one time only and used by means of multiple parts at various points in the system. Although the captioning of forms and the sequential arrangement of their itemization and arrangement of copies for distribution are functions of the analyst, he usually can depend on the sales representative of a reputable business forms and systems house to furnish a large measure of assistance. The representative acts as a catalyst by again thinking through the purpose and arrangement of the form with the analyst, often suggesting improvements and refining phrases and captions. The representative also accepts responsibility for such mechanical features as the substance and quality of the paper stock in relation to whether the form will receive a large volume of use, be filed, or be discarded; the best copy duplicating method in relation to the machine used in executing the form; the use of single or continuous forms in relation to frequency and method of preparation; and most important the sales representative will see to the drafting of the form and its proper design for efficient preparation and use by the methods and machines employed by the library.[4]

ASPECTS OF DESIGN—THE COMPUTER-BASED SYSTEM

The principles of design discussed in this chapter apply in the design of a computer-based system, varying only in the precision, logic, clarity, standardization, and specificity of information to be acted on and the anticipated

[4] For an epitomizing treatment of forms analysis and design discussed extensively in management literature see Systems and Procedures Association, *Business Systems,* The Association, Cleveland, Ohio, 1966, Chapter 10, pp. 1-61.

decisions to be implemented by the computer. The following outline indicates the areas with which the analyst should be concerned in designing a computer-based system.

1. *Characteristics of Available Computer System.*[5] Most libraries cannot justify the cost of maintaining a computer system for their exclusive use. Therefore it can be anticipated that most libraries will be faced with designing their operating systems to fit the capabilities of available outside equipment. The characteristics and availability of such equipment should be known:

(a) Means of input (punched cards, paper tape, telecommunication).
(b) Means of storage (magnetic tape, disk, drum, core memory).
(c) Storage capacity.
(d) Means of output (printout, punched cards, magnetic tape, cathode-ray tube).
(e) Relative speed of machine components.
(f) Computer time available for library operations.
(g) Available "off-line" equipment (sorter, tabulator, interpreter).

2. *Generalized Systems Chart.* A diagram showing the following should be made:

(a) The primary input and output.
(b) The major activities of the system.
(c) The interdependence among the various systems.

3. *Systems Requirements Analysis.*

(a) Analyze requirements in terms of a computer-oriented system, keeping in mind that through programming the computer system can be assigned many of the routine decisions found at all levels of management.
(b) Analyze for all possible applications of the principle of management by exception, determining and designing reporting systems best serving management's needs.

4. *Inputs/Outputs*

(a) Determine the input/output information required to satisfy the system, taking advantage of the speed of the computer.

[5] For information about various computers and their peripheral equipment see *Auerbach Standard EDP Reports,* a loose-leaf compilation "designed to satisfy the need for accurate, effective data to aid in the selection and utilization of computer systems . . .", Arthur H. Phillips, *Computer Peripherals and Typesetting: a study of the man-machine interface incorporating a survey of computer peripherals . . .* HMSO, London, 1968; and, Stanford L. Optner, *Systems Analysis for Business Management,* 2nd ed., Prentice-Hall, Englewood Cliffs, N. J., 1968, Chap. 7 Characteristics of Computer Equipment, pp. 124-150.

(b) Evaluate inputs/outputs of the present system and determine what additional inputs/outputs are required and/or what inputs/outputs are not required; determine the minimum essential information.

5. *Input Media*

(a) Evaluate existing forms or design new ones as required to capture needed data.

(b) Determine within the characteristics of available computer equipment the machine-readable media best adapted to the system's requirements (punched cards, paper tape).

(c) Evaluate and select equipment to be used in the library to generate machine-readable input records (key punch, tape typewriter).

6. *Computer Storage and Output Media*

(a) Determine media best suited for computer reports—for example, punched card, magnetic tape, printout, microfilm, within the limitations of available equipment.

(b) Determine for each record the type of computer storage media required for efficient processing (sequential or random) within the capabilities of available equipment, such as magnetic tape, data cell, disk, or drum storage.

(c) Evaluate and select any equipment to be used in the library to store and process computer output to yield records and reports, such as an electronic accounting machine for punched cards.

7. *Functions*

(a) Identify each function required for successfully meeting the system's requirements.

(b) Determine those functions to be performed under programmed control and those under manual control.

(c) Detail the procedures to be followed in the performance of each function.

8. *Decisions*

(a) Identify each decision required for successfully meeting the system's requirements.

(b) Identify those decisions that may be delegated to the computer under programmed control; if a decision involves personal intervention, identify the responsible person.

(c) Identify the source records and reports from which the information is obtained for making each decision.

9. *Resulting Actions*

(a) Identify the actions to be taken as the result of each decision.

(b) Identify the functions or further decisions resulting from each action.

The study staff should prepare the flow chart of the new system in sufficient detail (a) to furnish guidance for the programming of the computer-based parts of the system and (b) to permit preparation of detailed instructions for the manual procedures of the system. The study staff should continually modify the computer-based system until the system's objectives are realized. An acquisitions system is described in Chapter 9 to illustrate major aspects of computer-based systems design principles.

The final steps in the design of a computer-based or noncomputer-based system are to:

1. Estimate the cost of installing the new system and its operating cost.
2. Compare the cost of the old system with that of the new.
3. Determine the staffing requirements of the new system.
4. Prepare a summary report for the library's management outlining the features of the new system in relation to the old.
5. Prepare a detailed procedural manual and time schedule for conversion to and installation of the new system.

SYSTEMS INSTALLATION

The responsibility of the analyst should not end with the completion of the design phase of the systems study. The installation of the newly designed system should also be done by the systems analyst in cooperation with the line supervisor. It is logical that the installation should be the responsibility of the analyst since he designed the system, is aware of the nuances of his design, and has a definite interest in seeing that the system starts up successfully. The analyst should coordinate and correlate the various factors entering into the implementing of a new system—such as staff training, new forms, new equipment, and new facilities—so the installation can proceed according to a realistic time schedule. For example, the installation of a computer-based circulation system could involve coordinating the preparation and procurement of machine-readable identification cards, punched cards for books, data collection devices, computer programs, and new computer-produced forms such as overdue and recall notices.

There are three possible ways in which a system may be installed: (a) all at once on a crash program basis with little, if any, break in service during the transition; (b) *simultaneously* as a parallel program with both the old and the new system being concurrent for a predetermined time;

and (c) *step by step* on a piecemeal basis where portions of the new system are implemented within the old system. The size and nature of the new system will have considerable bearing on what installation approach the analyst may take. For example, it may be possible to implement a punched-card circulation system step by step. The preparation and insertion of the punched cards in the books may be accomplished on a piecemeal basis; that is, as books circulate punched cards could be made up and then be inserted when the books are returned. When a sizable portion of the active collection has been covered, the switch over to the punched card system could then be made. The simultaneous type of installation, although expensive, may be feasible where assurance that a new system will perform as accurately as the old is desired. This would be particularly true in accounting operations where the new system's tallies should correspond with those produced under the old system. The all-at-once approach may be used if a new operation is not considered complex, such as the use of photocopying machines to reproduce catalog cards. The program could be operational almost immediately after the card stock is available, the copier installed, and the staff trained.

SYSTEMS FOLLOW-UP

After installing the system the analyst should plan to check back on its progress. This is essential if he is to obtain on-site feedback about the system so that he in cooperation with the line supervisor can make any necessary revisions and adjustments in operations, forms, and staffing. The analyst's follow-up report may also serve to justify and increase management's confidence in the new system.

The appropriate time to make the follow-up survey will vary with the situation. In any event it may be well to audit the system a month after it has been declared operational. After this initial observation, it also may be well to have a review of the system annually. The follow-up report should not be done at a distance but by visiting the system and speaking with the staff and checking whatever controls may have been built into the system. For example, the statistics being kept in the new system may have been designed to reflect the effectiveness of the new system over the old. Reports on costs, unit production, and usefulness of equipment may be of value in the follow-up evaluation. Another method of testing quality control would be for the analyst to discreetly observe the progress of units through the system.

The changes and recommendations resulting from a systematic follow-up study should assure the continued success of the new system.[6]

APPENDIX I

Handbook of Manual Procedures—Acquisitions System

I. Preorder Search Subsystem
(Refer to Figures 8-1a through 8-1f for pictured progress of preorder search subsystem.)

[*Note:* The following instructions are pictured in Figure 8-1a-f]

A. Book request received (see Figure 8-2.)
1. Check all given information provided on the request slip for completeness and spelling. If complete or can be readily completed, do so and continue; if not, return to requestor.
2. Post record of the number of requests received on daily statistics sheets according to originating department.

B. Is the request for a series? If yes, see No. 1 of this item B; if not, see No. 2 of this item B.
1. If the request is for a series or part of a series (that is, a publication issued in successive parts and generally intended to be continued indefinitely), check the continuation card file in the acquisitions system and, if necessary, the central serials file in the serials system.
(a) If the library receives the series, record the request as a duplication, indicating the call number; check the STANDING/ORDER box and return to the sender.
(b) If not located, continue search in public catalog.
2. Search the public catalog
[The Dewey decimal catalog, for material before 1966, and LC (Library of Congress) catalog, beginning in 1966, are searched by main entry (author line) and/or by title, whichever the searcher feels is the more accurate approach. Place a check mark on the book purchase request form against the step taken, for example,

[6] For a thorough presentation of the subject of installing new procedures and follow-up see Richard F. Neuschel, *Management by System,* 2nd ed., McGraw-Hill, New York, 1960, Chap. 16, Installing Approved Procedures, pp. 298-322.

check in boxes, OFF. CAT., PROOF CD, PUB. CAT., and so forth.]

[*Note:* The following instructions are pictured in Figures 8-1b and top of 8-1c.]

(a) Is the book already in the library collection?

1. If the book is found to be in the collection and the request is not for an added copy, the call number is recorded on the request slip in the box, IN LIB LOC & CALL NO. The LIB HAS box is checked and the appropriate box on the side of the request slip is checked (PUB. CAT.). The request form is recorded on the daily statistics as a duplication by department and is returned to the requestor.
2. If the request is for a departmental library and the book is found to be in that library, the DEPT HAS box is checked, the call number is recorded in the IN LIB LOC & CALL NO. box, and the request is recorded as a duplicate and returned to the requestor.
3. If the book is found in the collection and the request is for another copy for the general library or a departmental library, the call number is entered in the box, IN LIB LOC & CALL NO. In the LC CARD NO. box, "AC" (Added Copy) is recorded to indicate there is no need to order Library of Congress cards. The following information is verified and/or recorded on the request slip:
   (a) Official main entry (author line)
   (b) Publication date
   (c) Publisher and series note
   The appropriate boxes on the request slip are checked and the search is continued in the computer-produced "In Process" list, and, in *Books in Print* (BIP), the latter for ascertaining the price and the publisher.

(b) If the book is not in the library, the search is continued.

C. Search title in *In Process* List

(This computer-produced list contains books ordered and not received as well as books ordered and received but not cataloged. It is arranged alphabetically by title.)

1. If the title being searched is found in this list and is not a request for an added copy, it is considered a duplication and recorded on the statistics sheet by department, the ON ORDER box is checked, and the request slip is returned to the originator.
   (a) If the request is for an added copy, fill in box IN LIB LOC & CALL NO., enter "AC" in LC CARD NO. box, verify or correct author and title, and continue search in BIP.
2. If title is not found, continue search.

D. Search title in Library of Congress proof-card file (file is alphabetical by title and located in the acquisitions system).
   1. If appropriate card is found, take from file:
      (a) If necessary, correct main entry (author entry) on the request slip to conform with proof card.
      (b) If main entry on request slip differs from proof card, recheck the public catalog and in process list if the search was made under a different entry. Also check the continuation file and, if necessary, the central serials file if a series note is now evident.
      (c) Record LC card number and write "proof" in the LC CARD NO. box, to indicate that there is no need to order LC cards.
      (d) Attach the LC proof card to request slip and continue the search for the price of the book.
   2. If no LC proof card is located, continue search.

E. Search *Books in Print* (BIP bound cumulative volumes issued annually; approach is by subject, author, or title; this is an index to most domestic publishers' lists).
   1. If entry is found, verify and/or record the following:
      (a) Price
      (b) Publication date
      (c) Publisher

      If all information necessary for ordering the book (namely, the library does not have, or have on order, or in process, and the author, title, LC card number, price, publisher, and publication date) is recorded on the request slip, the search is completed; give to the acquisitions librarian.

[*Note:* The following instructions are pictured in Figures 8-1c and top of 8-1d.]

F. Search Forthcoming Books (FORTH. BOOKS) (Bimonthly publication allowing approach by author and/or title. Publication date to the month is included in entry.)
   1. If entry is found, verify and/or record the following information:
      (a) Price
      (b) Publication date
      (c) Publisher

      With this information recorded, forward request to acquisitions librarian.
   2. If not found, continue search.

G. Search request in domestic or foreign publishers' catalogs (PUBL. CAT.) and/or *Publisher's Trade List Annual* (PTLA is an annual bound compilation of American publishers' catalogs).
   1. If located, verify and/or record the following information:
      (a) Price

(b) Publication date
(c) Publisher

If all information needed to order is complete, forward to acquisitions librarian.

2. If not found, continue search.

[*Note:* The following instructions are pictured in Figures 8-1d and top of 8-1e.]

H. Search *Publishers' Weekly* (PW; a weekly listing by author of new books) and/or *Book Publishers' Record* (BPR; a monthly cumulation of PW lists, classified by Dewey decimal numbers, with an author, title, and subject index).
1. If located, verify and/or record the following information:
   (a) Price
   (b) Publication date
   (c) Publisher
2. If not found, continue search.

I. Search request in *Cumulative Book Index* [CBI; a cumulative world list of books in the English language; an author, title, or subject approach is possible; the LC card number is usually listed under the main (author) entry].
1. If located, verify and/or record:
   (a) Main entry
   (b) LC card number
   (c) Publisher
   (d) Publication date
   (e) Price
   (f) Series note (If a new or different series note is discovered, check the continuation file and, if necessary, the central serials file.)

   With this information recorded, forward to acquisitions librarian.
2. If the main entry is different from that searched, begin search over again in public catalog.
3. If not located, continue search.

[*Note:* The following instructions are pictured in Figures 8-1e and 8-1f.]

J. Search in *National Union Catalog* (NUC; an extensive set of volumes issued in various editions, containing retrospective cataloging data on most books and series to be found in the Library of Congress and other large research libraries; approach is by official main entry).
1. If found, verify and/or record the following information:
   (a) Publisher
   (b) Publication date
   (c) LC card number

(d) Series note (If series note is found, check the continuation file and, if necessary, the central serials file.)

With this information recorded, forward request to acquisitions librarian.

2. If not located, place "XXX" in the LC CARD NO. box. Forward to acquisitions librarian.

K. Request reviewed by acquisitions librarian
   1. Is there enough information to purchase book?
      (a) If yes, send request to order subsystem.
      (b) If no, either return to requestor for additional information or to searcher for suggested alternative approaches and sources.
   2. Is the book apparently out of print?
      (a) If not, send request to order subsystem.
      (b) If out of print (OP), notify requestor and refer request slip to desiderata file in order subsystem.

FIGURE 8-1a-f Flow-chart of manual procedures acquisitions system—preorder search subsystem

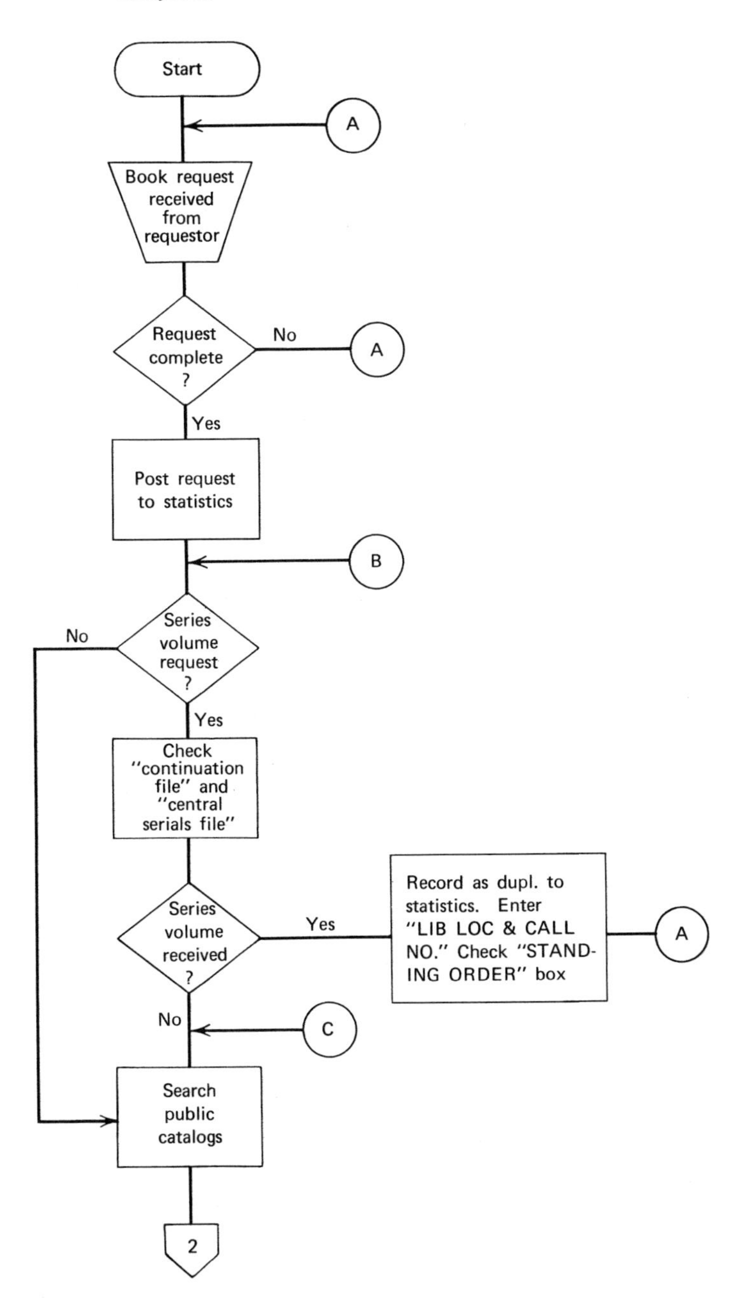

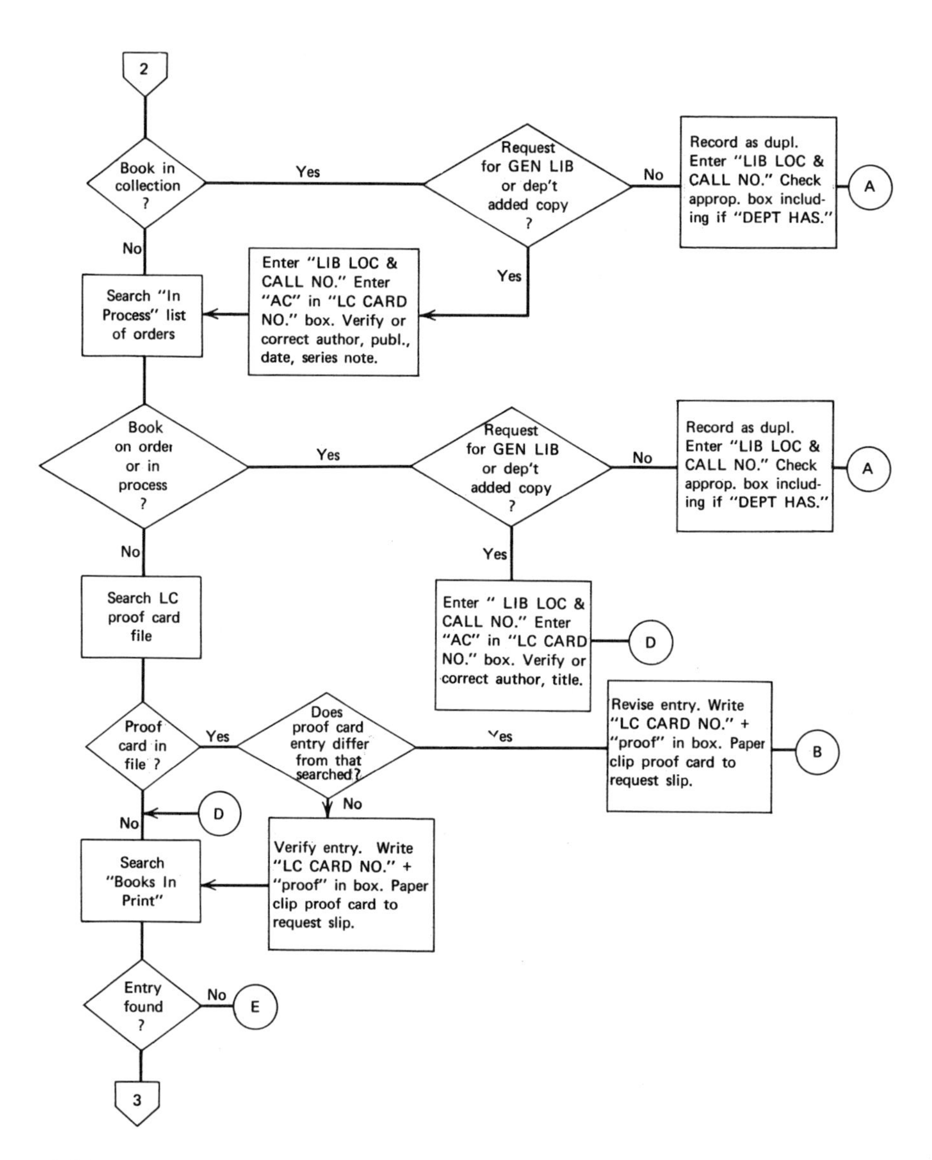
2
Book in collection ?
Yes
Request for GEN LIB or dep't added copy ?
No
Record as dupl. Enter "LIB LOC & CALL NO." Check approp. box includ-ing if "DEPT HAS."
A
No
Yes
Search "In Process" list of orders
Enter "LIB LOC & CALL NO." Enter "AC" in "LC CARD NO." box. Verify or correct author, publ., date, series note.
Book on order or in process ?
Yes
Request for GEN LIB or dep't added copy ?
No
Record as dupl. Enter "LIB LOC & CALL NO." Check approp. box includ-ing if "DEPT HAS."
A
No
Yes
Search LC proof card file
Enter " LIB LOC & CALL NO." Enter "AC" in "LC CARD NO." box. Verify or correct author, title.
D
Proof card in file ?
Yes
Does proof card entry differ from that searched?
Yes
Revise entry. Write "LC CARD NO." + "proof" in box. Paper clip proof card to request slip.
B
No
D
No
Search "Books In Print"
Verify entry. Write "LC CARD NO." + "proof" in box. Paper clip proof card to request slip.
Entry found ?
No
E
3

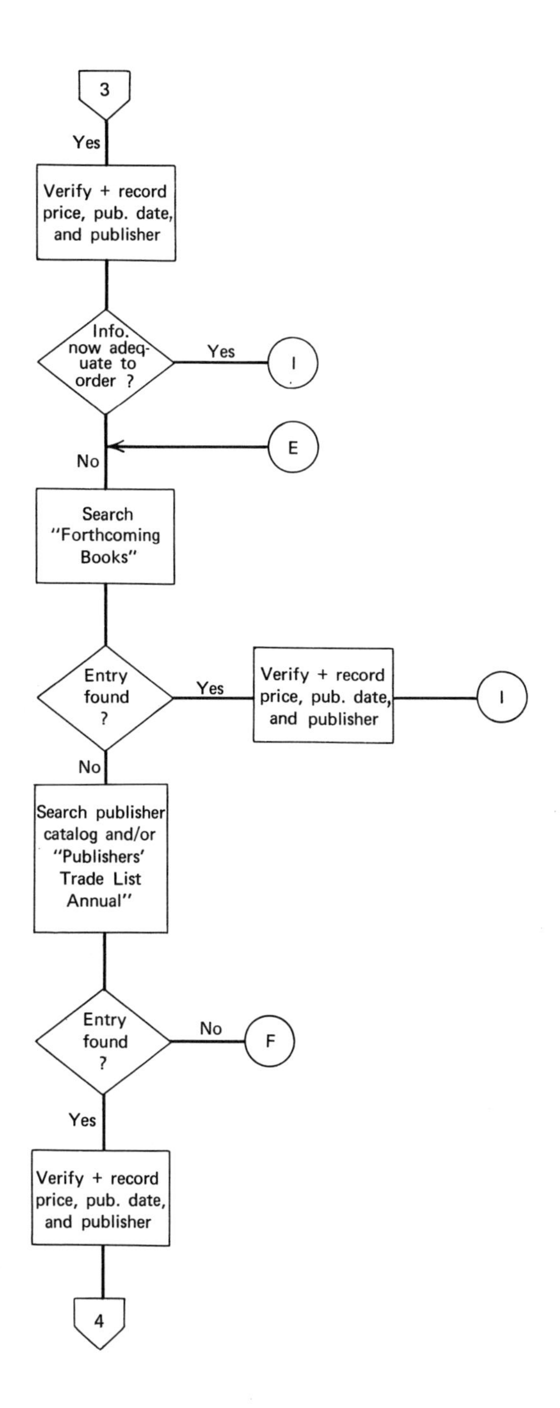
3
Yes
Verify + record price, pub. date, and publisher
Info. now adeq-uate to order ?
Yes
I
E
No
Search "Forthcoming Books"
Entry found ?
Yes
Verify + record price, pub. date, and publisher
I
No
Search publisher catalog and/or "Publishers' Trade List Annual"
Entry found ?
No
F
Yes
Verify + record price, pub. date, and publisher
4

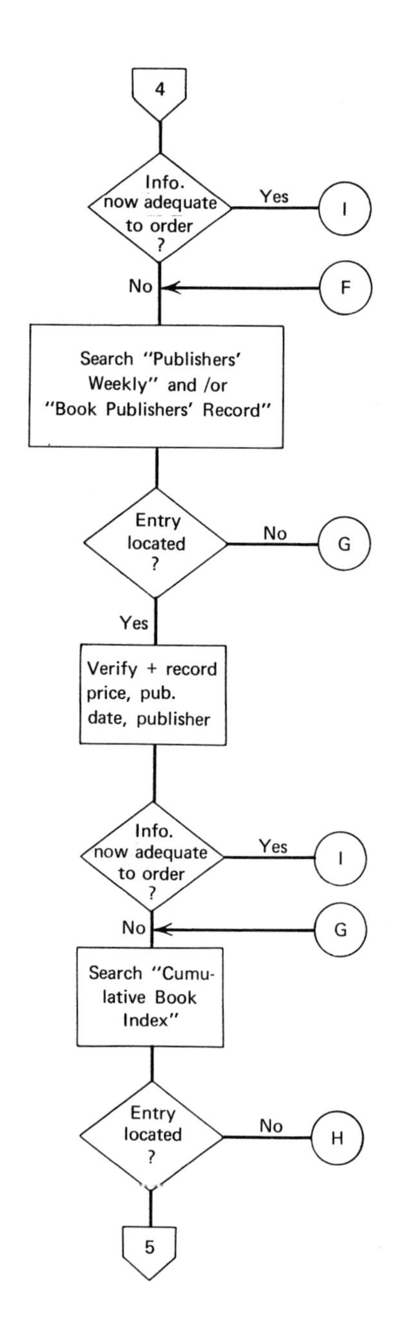
4
Info.
now adequate
to order
?
Yes
I
No
F
Search "Publishers' Weekly" and /or "Book Publishers' Record"
Entry
located
?
No
G
Yes
Verify + record price, pub. date, publisher
Info.
now adequate
to order
?
Yes
I
No
G
Search "Cumulative Book Index"
Entry
located
?
No
H
5

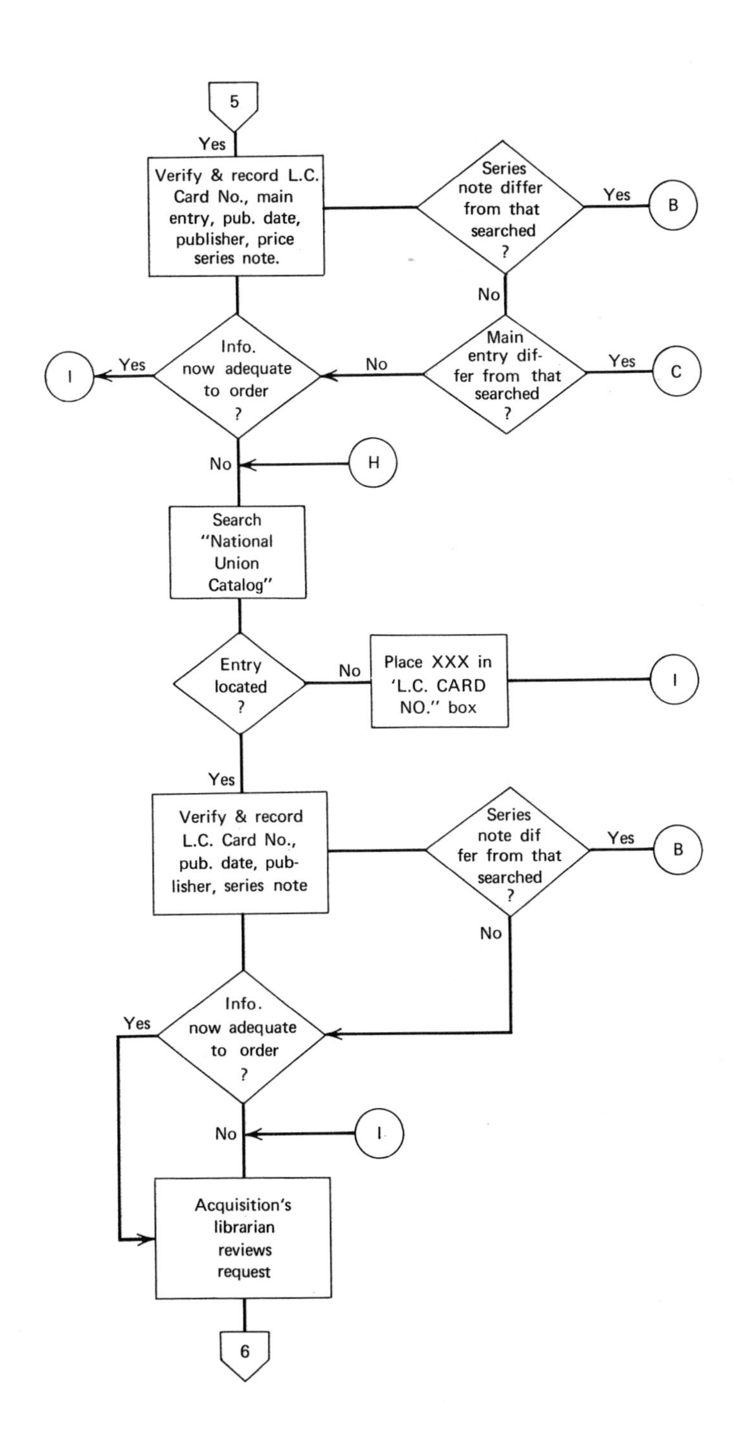
5
Yes
Verify & record L.C. Card No., main entry, pub. date, publisher, price series note.
Series note differ from that searched ?
Yes
B
No
Info. now adequate to order ?
Yes
I
No
Main entry differ from that searched ?
Yes
C
No
H
Search "National Union Catalog"
Entry located ?
No
Place XXX in 'L.C. CARD NO." box
I
Yes
Verify & record L.C. Card No., pub. date, publisher, series note
Series note differ from that searched ?
Yes
B
No
Info. now adequate to order ?
Yes
No
I
Acquisition's librarian reviews request
6

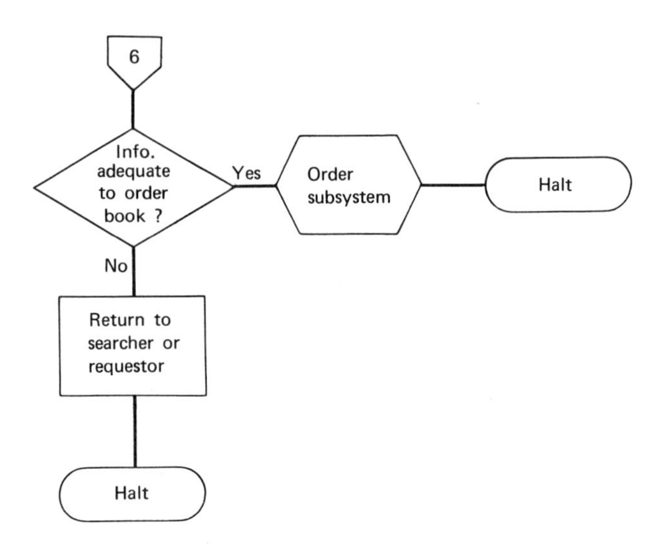

FIGURE 8-2 Book purchase request

BOOK PURCHASE REQUEST
RENSSELAER POLYTECHNIC INSTITUTE

ORDER NO.	FUND NO.	DEPT.

CLASS RESERVE DATE REQUIRED	EDITION	PUB. DATE	NO. OF COPIES	L. C. CARD NO.	DATE REQUESTED

AUTHOR:

TITLE

PUBLISHER

ADDRESS OF PUBLISHER

VENDOR

NOTES:

L-74 ☐ GIFT ☐ GP: ______ ☐ CAT. SEP.

EST. PRICE

☐ LIB. HAS
☐ DEPT. HAS
☐ ON ORDER
☐ ON RESERVE
☐ STANDING ORD.

IN LIBRARY LOC. & CALL NO.

OFF. CAT.
PROOF CD.
PUB. CAT.
B. I.P.
C. B. I.
FORTH. BOOKS
PUBL. CAT.
N. U. C.
OTHER
CITED IN

REQUESTED FOR
☐ UNDERGRADUATE
☐ GRADUATE
☐ RESEARCH
☐ FACULTY
☐ OTHER

REQUESTED BY

APPROVED BY

APPENDIX II

Manual of Operations—General Library, Rensselaer Polytechnic Institute

INTRODUCTION AND OUTLINE

I. Purpose

The *Manual of Operations for the Rensselaer General Library* provides:

A. Description of the current organizational structure of the library

B. Outlined statements of the programs, policies, procedures, and instructions governing the operations and the administration of the library.

II. Outline of Manual

This manual is divided into five major sections based primarily on the library's organizational structure as of Spring 1969 (see Figure 8-3). Each section of the manual has been assigned to a specific administrative department or division except for the 0000 section which is library-wide. The outline is as follows:

A. 0000–0999 Programs, Policies and Interpretations (library-wide)

0000	Table of Contents
0001–0019	Organization of the Library
0020–0299	Administrative Policies
0300–0399	Fiscal Policies and Procedures
0400–0499	Housekeeping Policies and Procedures
0500–0599	Personnel Policies and Procedures
0998	Schedule for Maintenance, Storage and Disposition of Records
0999	Forms

B. 1000–1999 Administration and Planning Office

1000	Table of Contents
1001–1099	Policies and Procedures
1100–1199	Administrative Services Section Policies and Procedures
1500	Schedule for Maintenance, Storage and Disposition of Records
1501	Forms

C. 2000–2499 Technical Services Division

2000	Table of Contents

FIGURE 8-3 Administrative structure chart, Spring 1969. Rensselaer Polytechnic Institute—General Library

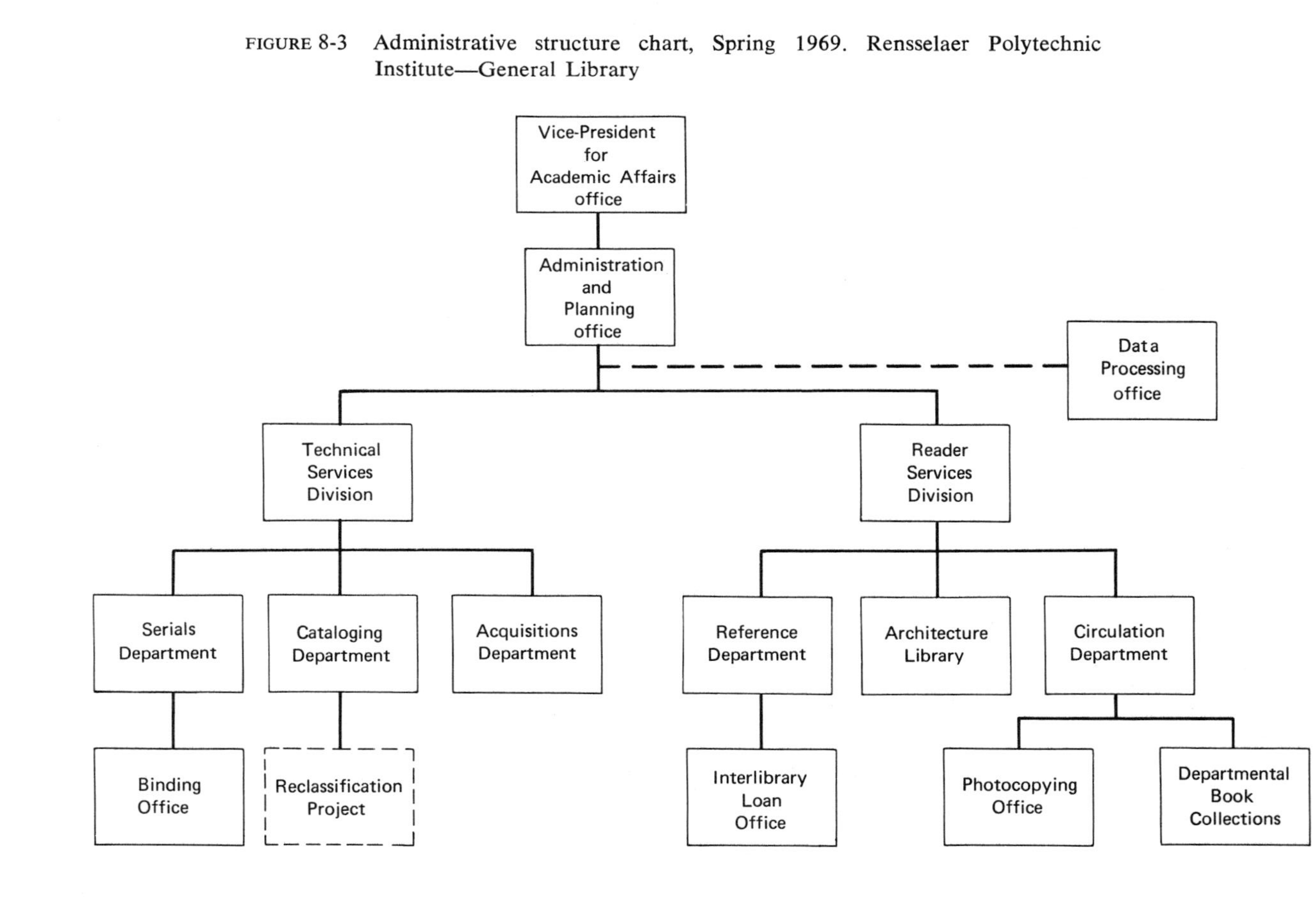

2001–2097 Policies and Procedures
2098 Schedule for Maintenance, Storage, and Disposition of Records
2100–2199 Serials Department Program, Policies, and Procedures
2200–2249 Binding Office Programs, Policies, and Procedures
2250–2349 Cataloging Department Programs, Policies and Procedures
2350–2399 Reclassification Project Programs, Policies, and Procedures
2400–2497 Acquisitions Department Programs, Policies, and Procedures
2498 Schedule for Maintenance, Storage, and Disposition of Records
2499 Forms

D. 2500–2999 Reader Services Division
2500 Table of Contents
2501–2597 Policies and Procedures
2598 Schedule for Maintenance, Storage, and Disposition of Records
2599 Forms
2600–2699 Reference Department Programs, Policies, and Procedures
2700–2749 Interlibrary Loan Office Programs, Policies, and Procedures
2750–2799 Architecture Library Programs, Policies, and Procedures
2800–2899 Circulation Department Programs, Policies, and Procedures
2900–2949 Photocopying Office Programs, Policies, and Procedures
2950–2997 Departmental Book Collections Programs, Policies, and Procedures
2998 Schedule for Maintenance, Storage, and Disposition of Records
2999 Forms

E. 3000–3499 Data Processing Office
3000 Table of Contents
3001–3497 Programs, Policies, and Procedures
3498 Schedule of Maintenance, Storage, and Disposition of Records
3499 Forms

SECTION 2900

Photocopying Office Programs, Policies, and Procedures for Processing Photocopying Office Charge Accountss

Program: University staff, faculty, and graduate researchers are able to charge journal and other photocopying costs to their research or department's account fund number and be billed on a monthly basis through the university's "transfer of funds" procedure.

Policy: The photocopying office of the General Library provides a charging service to the university community and, as a result, must prepare, mail out, and submit to the university's comptroller authorized transfers of funds. Of primary importance within the billing operation is the maintenance of journal-use statistics for collection evaluation and journal-use measurement.

Procedure: The following is the procedure for processing charge accounts to produce the transfers of funds and journal-use statistics within the photocopying office:

Photocopying Staff	*Time*	*Procedure*
Library clerk or student assistant	As received	1. Receive photocopy request from library users (Form 2999–1, Photocopy Request Form). 2. Reviews the request form for completeness. 3. If the request form is incomplete, return for completion. If the request form is complete, accept materials to be photocopied and place on "work shelf" with request form.
Library clerk or student assistant	Within 8 hours	1. Take photocopy requests and journals from work shelf and photocopy according to directions on the request form. 2. Photocopying completed, sign log for number of copies done, collate, paper clip and place copies in 8½ × 11 envelope. Attach photocopy request form to outside of envelope. 3. Compute price of photocopying at 7 cents a page and write it in total cost section of the request form, item 5. 4. Place completed job in appropriate

Photocopying Staff	*Time*	*Procedure*
		alphabetical slot under counter in office.
Library clerk or student assistant	As requested	1. When user calls for his copies by name, give him his envelope and review the request form. If not already done, have him write in the account number and his name. 2. Place request form in out basket.
Library clerk	5 P.M.	1. Give request forms from out basket to evening student assistants at the circulation desk for coding of the request forms for keypunching operation.
Evening student assistant	6 P.M.-11 P.M.	1. Encode request form by circling the valid data to be keypunched where a charge is made (refer to Form 2999-1). 2. Abbreviate journal titles according to Form 2999-2, Word Abbreviation List. . . . 3. Place completed forms on circulation librarian's desk.
Circulation librarian	As received	1. Review and edit any difficult unabbreviated journal titles. 2. Give work to keypunch operator.
Keypunch operator	As received (usually 4-5 P.M.)	1. Insert program card (Form 2999-3) and prepare punched cards (Form 2999-4) for request forms. 2. File request forms in monthly file. 3. Place punched cards in the month's file located in the circulation librarian's office.
Keypunch operator	First of the month	1. Take month's punched cards to computer center. Leave with computer operator for processing.
Computer operator	As received	1. Using program LIBRC 8 (Xerox Accounting Program) computer produces error listing (Form 2999-5) as phase one.

Photocopying Staff	*Time*	*Procedure*
		2. Give error listing to library keypunch operator.
Keypunch operator	As received	1. Correct error listing by punching new cards for monthly file. 2. Return month's deck of cards to computer operator.
Computer operator	As received	1. Run through phases two and three of the LIBRC 8 to produce the transfers of funds (Form 2999-6 and 2999-6a) in triplicate and to update the monthly statistics tape. 2. Run LIBRC 9 (Xerox Statistics Program) to print out requested statistical tables.
Keypunch operator and circulation librarian	By first week of every month	1. After picking up transfers of funds, decollate and cut to appropriate size (8½ × 11). 2. Review transfers of funds for any obvious errors. 3. If errors are found, issue a manually prepared correct transfer of funds and pull faulty cards from punched card deck. 4. If no errors, keep office copy and mail two copies to department chairman for signature authorizing transfer of funds.
Circulation librarian	As required	1. Match authorized transfer of funds against office copy.
Circulation librarian	25th of each month	1. Send authorized transfers of funds to Comptroller's office for processing. 2. Include in mailing the month's deck of punched cards for their EAM operations.
Circulation librarian and head, reader services division	As required	1. Review statistical data on journal use. Evaluate retention policies and compact storage for journals apparently used only infrequently. Study journal use by subject, year, and department and status of user.

SECTION 2998

Schedule for Maintenance, Storage, and Disposition of Records

Form Title and/or Number	*Storage and Disposition*
Photocopy Request Form (2999-1)	Maintain two months after date form was filled out.
Abbreviations (2999-2)	Maintain as long as list is current. Discard as revisions are received.
Program Card (2999-3)	Maintain as long as usable. Duplicate and discard original when worn.
Punched Card (2999-4)	Monthly, send to Comptroller's office with authorized transfer of funds.
Error Listing (2999-5)	Discard as corrected.
Transfer of Funds (2999-6), (2999-6a)	Maintain office copy to match with authorized transfer of funds received from departments. Send authorized copy to Comptroller. Keep office copy one year.

Form 2999–1

RENSSELAER POLYTECHNIC INSTITUTE GENERAL LIBRARY
PHOTOCOPY REQUEST FORM

PLEASE PRINT

Journal name ______________________ First line of call no. ____________
7 8

Volume/Issue no. ____________ Year ____________ Month ____________
9

Abbreviated Article Title ______________________________________

______________________________ Author ____________________

Paging ____________________ No. of copies of each page ____________

Special Instructions ______________________________________

Journal name ______________________ First line of call no. ____________
7 8

Volume/Issue no. ______________ Year ____________ Month ____________
9

Abbreviated Article Title ______________________________________

______________________________ Author ____________________

Paging ____________________ No. of copies of each page ____________

Special Instructions ______________________________________

Academic Status:
4

__a. Faculty __c. Undergraduate
__b. Graduate __d. Other

Dept. ______________________
3

Date ______________________
1

Print name ______________________
2

____

Cash____ Charge____

CHARGE

__________ pages @ __________¢

Total cost $__________
5

Project #__________
6

(11–68rev.5M)

Form 2999–2

(Title page of the 33-page list included in Manual in its entirety)

UNITED STATES OF AMERICA STANDARDS INSTITUTE

Sectional Committee Z39 on Library Work and Documentation:

Subcommittee on Periodical Title Abbreviations

Revised and Enlarged Word-Abbreviation List

for

USASI Z39.5-1963

American Standard for Periodical Title Abbreviations

Published by the

National Clearinghouse for Periodical Title Word Abbreviations

c/o Chemical Abstracts Service
University Post Office
Columbus, Ohio 43210

December 31, 1966

Form 2999–3

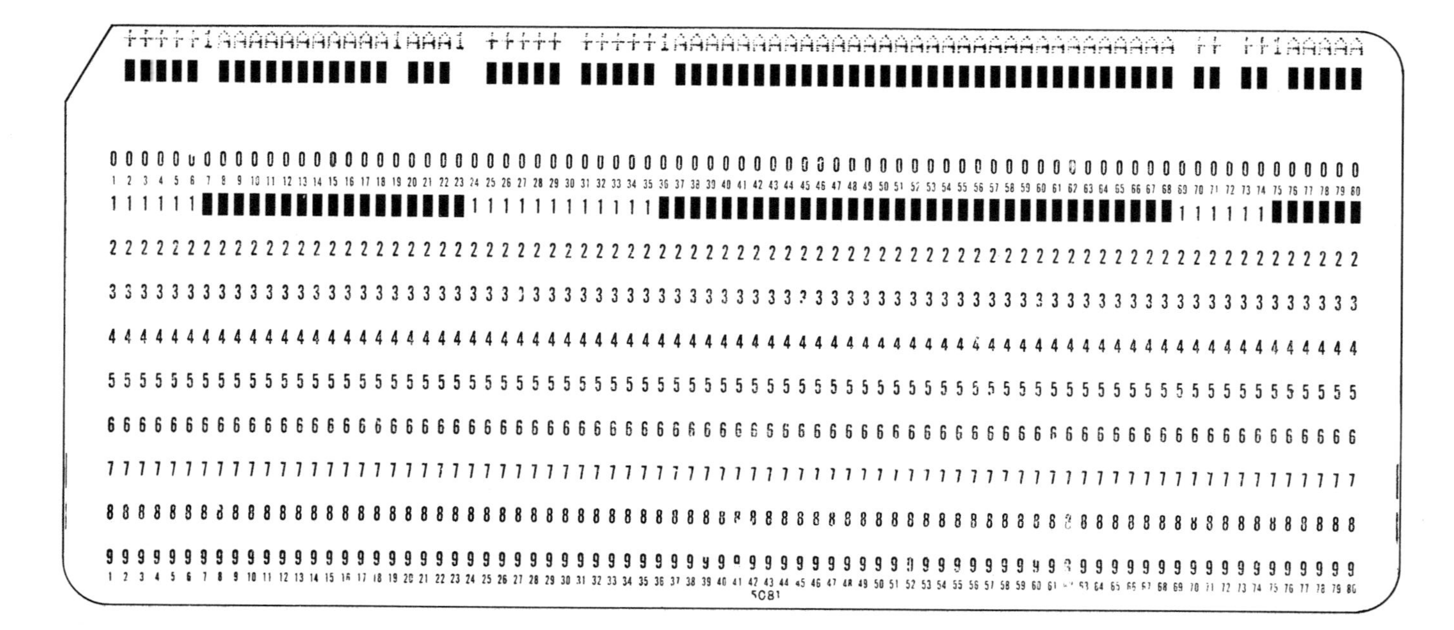

Form 2999–4

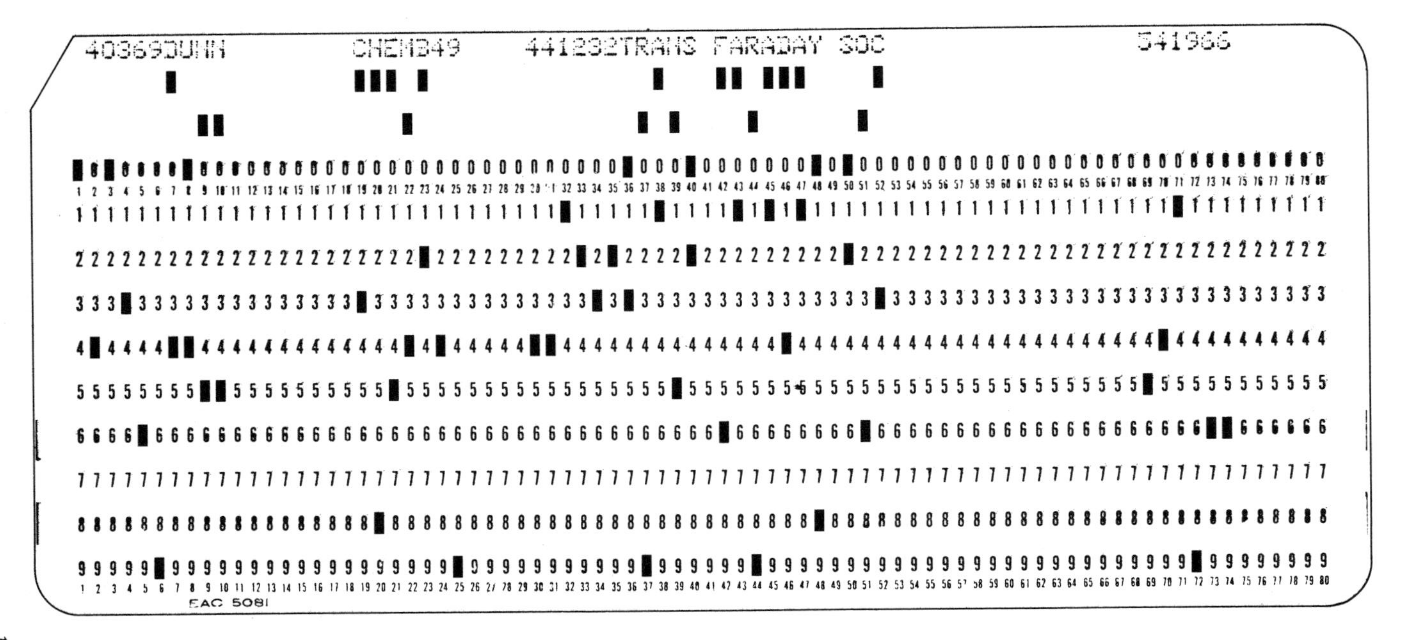

40369DUNN CHEM849 441232TRANS FARADAY SOC 541966
EAC 5081

Form 2999–5

Form 2999-5

Error Listing

```
XEROX ACCOUNTING AND STATISTICS PROGRAM
LEGEND FOR PHASE 1  -- SYNTAX CHECK
$ FLAGS THE LOCATION OF ERROR,CODE MEANS-
1. THIS CARD IS A THIRD CONTINUATION
2. INVALID CHARACTER OR IMPOSSIBLE MM OR DD IN DATE
3. DEPARTMENT DOES NOT EXIST
4. INVALID STATUS -- NOT A THRU D
5. AMOUNT CHARGED HAS NON - NUMERIC CHARACTERS
6. AMOUNT CHARGED IS OVER $9.99 WITHOUT PERMISSION
7. PROJECT NUMBER IS NON - NUMERIC IN CONTENT
8. CALL NUMBER IS NON NUMERIC
9. YEAR IS NON - NUMERIC
      CARDIMAGE FLAGGED AND REASONS APPEAR BELOW

                         00000173
080169CHABILCAS        PHYSB42     447722BI
                                                                $8 $9
                         00000175
080469KAUFMAN          MS  B28     437082APPL MECH REV
                                                                $8 $9
                         00000555
082169BAILEY           CHEMA       441292C R
                                                                $8 $9
                         00000565
082169YONG             MS  B21     1022018
                                                                $8 $9
CC00004  ERRORS FLAGGED IN THIS RUN .
```

Form 2999-6—2999-6a

Form 2999-6

RENSSELAER POLYTECHNIC INSTITUTE

TROY, NEW YORK

M E M O R A N D U M

TO- FRANK BUCKLEY
COMPTROLLER

FROM- AUDREY K. DAVIS
ADMINISTRATIVE ASSISTANT

SUBJECT- TRANSFER OF FUNDS

WHEN SIGNED BY THE DEPARTMENT HEAD, THIS WILL AUTHORIZE THE TRANSFER OF FUNDS FROM THIS DEPARTMENT'S PROJECT ACCOUNT NUMBERS LISTED BELOW TO THE LIBRARY ACCOUNT NO. 190.294 FOR XEROX COPIES.

DEPARTMENT- MECHANICS DATE- JANUARY

DATE	PROJECT NUMBER	AMOUNT
01/07/68	754.07	$ 7.40
01/03/69	154.201	4.06
	154.201	1.33
	154.201	3.36
	154.201	5.74
	154.201	2.17
	154.201	3.22
	154.201	1.82
	154.201	1.82
01/08/69	454.44	.56
01/09/69	754.07	.84
01/15/69	154.201	.28
01/27/69	154.201	1.05
01/30/69	754.07	1.33
	754.07	.14

DEPARTMENT HEAD / DATE

AUDREY K. DAVIS / DATE

Form 2999-6a

ITEMIZED INDIVIDUAL TRANSFER OF FUNDS

PROJECT NUMBERS	NAMES OF CHARGERS	AMOUNT CHARGED
154.201	BIRNBOIM	$ 3.22
	BIRNBOIM	2.17
	BIRNBOIM	5.74
	BIRNBOIM	3.36
	BIRNBOIM	1.33
	BIRNBOIM	4.06
	BIRNBOIM	1.82
	BIRNBOIM	1.82
	OLENDER	1.05
	WALTER	.28
454.44	BURKE	.56
754.07	HAWLEY	.84
	HAWLEY	.14
	HAWLEY	1.33
	HAWLEY	7.40

SUM TOTAL OF TRANSFER OF FUNDS FOR MONTH OF JANUARY $35.12

chapter nine

SYSTEMS DESIGN—COMPUTER-BASED ACQUISITIONS SYSTEM

We shall now discuss a case study of an operational computer-based acquisitions system that applies the major principles and aspects of designing a system as outlined in the previous chapter. This case study is intended to illustrate the characteristics of the computer system used as well as the input/output media involved in satisfying the system's requirements. As with the following chapter on the design of computer-based operations for a serials system, critical comments and observations are added.

CHARACTERISTICS OF THE AVAILABLE COMPUTER SYSTEM[1]

As stated in Chapter 8 the capabilities and limitations of the computer system available to the library must be known prior to designing the computer-based library system. In the case of the described acquisitions system the equipment available to the library was an International Business Machines System 360/50 computer installation consisting of the following devices listed with their IBM type number and related capabilities.

[1] For explanation of unfamiliar computer terms, see Charles J. Sippl, *Computer Dictionary and Handbook,* Howard W. Sams, Indianapolis, Ind., 1966.

1. First there was a 2050 central processing unit (CPU) with type 1052 typewriter and console that could handle compilation and execution of programs together. CPU core storage is 131,172 bytes (characters). Two 16-digit decimals may be multiplied together in 37.25 microseconds. This speed of data handling is illustrated by the fact that 18 three-line bibliographic records (call number, author, and title) plus a page heading may be formatted for printing in less than 1 second including an edit (verification) of the call number and allowance for spaces between entries.

2. There were two 1403 line printers for printed output. Each printer is rated at 600 lines per minute with 132 characters per line using a printing chain with the Standard Character Set of 48 graphics (26 upper case alpha, 10 numeric, and 12 assorted special characters). Using the Full Character Set (Text Printing) chain, which includes upper and lower case letters, numerals, and many more special characters than the Standard Set, printing speed is reduced by more than 50 percent to about 250 lines per minute.

3. A 2540 card reader punch was available for reading punched-card input and punching output data. This machine reads 1000 cards per minute and punches 300 cards per minute.

4. There were four 2311 disk drives that house "compilers" and the various "monitor systems." Compilers are programs that translate instructions in specific programs into the language that the machine can use. Monitor systems coordinate and direct the computer system's capabilities for maximum effectiveness in processing user programs. There are 7.25 million characters of random (direct) access storage available from each disk drive and it may be read at 56,000 characters per second. Direct access is essential to such operations as sorting or putting a file in some desired order. For example, 3000 variable-length records can be alphabetized in less than 3 minutes including the time it takes for the input and output magnetic tapes to be rewound.

5. Five 2400 model 1 magnetic tape drives for general input/output (I/O) operations were available. There are 23 million characters of sequential storage available per tape drive deliverable at 30,000 characters per second. The tapes are useful for files that will be used but not rearranged in processing.

6. A 2671 paper-tape reader was provided. With the feed and take-up spools, paper tape is read at 1000 characters (about three book orders) per second. The spooling accessory allows tapes to be read in proper data sequence without respect to how the paper tape is wound.

7. A 2361 bulk core storage provided over one million bytes of core

storage supplementing the storage capacity available in the central processing unit.

8. A 2841 control unit for the disk 2311 drives and a 2821 control unit for the 1403 printers and the 2540 card reader were available to synchronize the I/O devices with the CPU so that the program processing would continue during the I/O operations. Being mechanical, these operations are considerably slower than the electronic operations of the CPU.

9. Peripheral utility type equipment was also available including the 129 keypunch, the burster for breaking apart paper printout, the decollator for pulling out carbon paper and separating multiple copies of a printout, the punched-card duplicating machine, and the punched-card sorter.

SYSTEMS CHART

In general the acquisitions system, as illustrated in Figure 9-1 which indicates the inputs/outputs and the major activities of the system, is viewed as the purchasing department for a primary input of the library—the book. The chart illustrates the effect of the user's request, the invoice, and the book on the interdependent subsystems of acquisitions.

REQUIREMENTS AND INPUT/OUTPUT MEDIA

When the decision to purchase a book was made, the analyst could see the effect this would have on the acquisition system as well as others. He applied the principle of economy of recording information only once and began the design that could satisfy this principle. He also kept in mind five major internal requirements that the new acquisitions system had to satisfy: a current on-order file, a claiming procedure, a notification to requestor, various fiscal reports, and operating statistics.

In order to satisfy the systems need for records in two forms, eye legible and machine readable, an input device was needed to create the records in these formats. The key punch, although it produces punched cards that are eye legible and machine readable, was unacceptable because it could not prepare the needed multiple-copy order sets. On the other hand the paper-tape typewriter, like a standard typewriter, produces legible copy as it simultaneously punches the same information into a paper tape that serves as input to the computer. The advantages of the tape typewriter over the key punch in this ordering application appear to be (a) the production of a conventional, typed record that facilitates the

FIGURE 9-1 Acquisitions systems chart

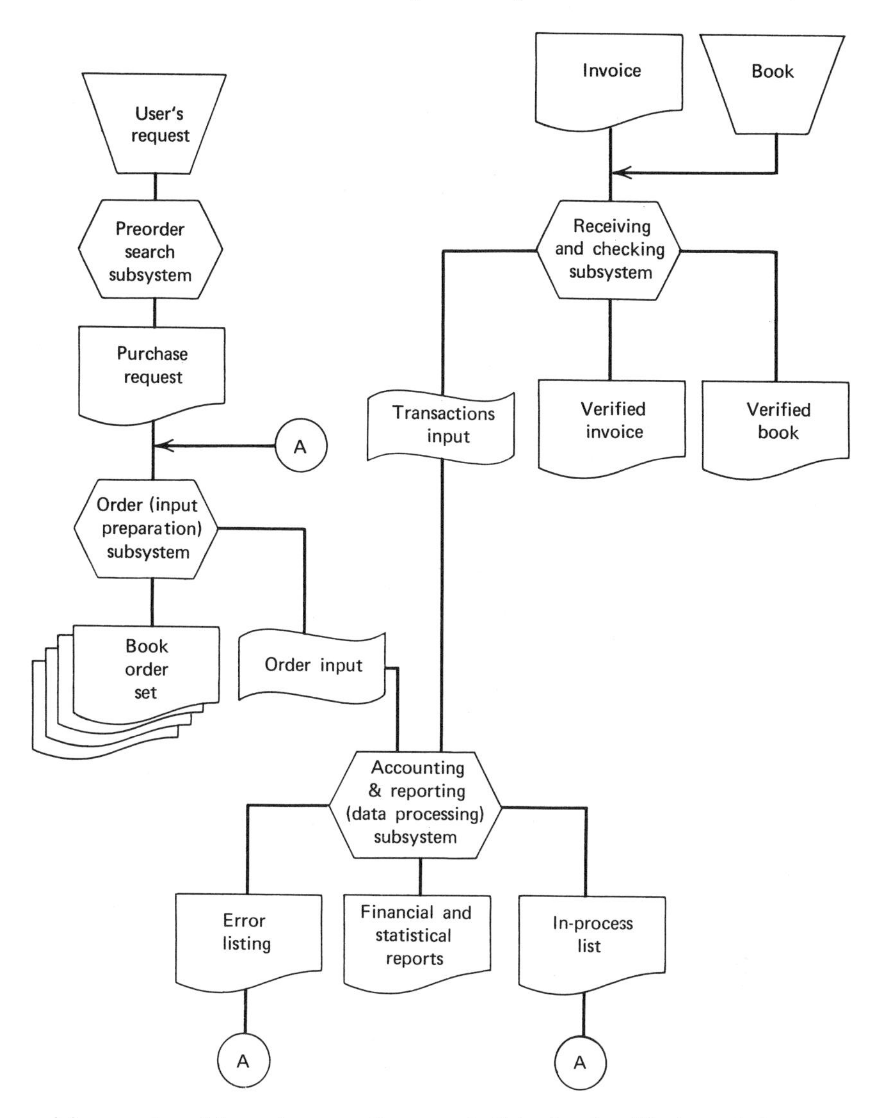

revision and making of corrections and (b) the familiarity of the typewriter as a tool to all library clerical personnel.

The tape typewriter used in this application is controlled by a master tape that performs the following functions:

1. Positions the order form and controls the horizontal and vertical positioning of data.

FIGURE 9-2 Multipart book order form

THIS BOOK ORDERED ON DATE INDICATED

DATE ORDERED	ORDER NO.	FUND NO.	D	ORIG. DEPT.	R	EDITION	PUB. DATE	NO. COPIES	L. C. CARD NO.

AUTHOR

(4)

TITLE

L. C. ORDER SLIP

DATE ORDERED	ORDER NO.	FUND NO.	D	ORIG. DEPT.	R	EDITION	PUB. DATE	NO. COPIES	L. C. CARD NO.

AUTHOR

(3)

TITLE

ACTION

REPORT SLIP - RETURN WITH BOOK OR REPORT ON REVERSE SIDE

DATE ORDERED	ORDER NO.	FUND NO.	D	ORIG. DEPT.	R	EDITION	PUB. DATE	NO. COPIES	L. C. CARD NO.

AUTHOR

(2)

TITLE

ORDER SLIP - SEE REVERSE SIDE

DATE ORDERED	ORDER NO.	FUND NO.	D	ORIG. DEPT.	R	EDITION	PUB. DATE	NO. COPIES	L. C. CARD NO.

AUTHOR

(1)

TITLE

PUBLISHER AND PLACE

VENDOR

EST. PRICE	ACTUAL COST	DATE RECEIVED

REC.

RENSSELAER POLYTECHNIC INSTITUTE
LIBRARY BOOK ORDER DEPT.
TROY, NEW YORK 12181

FIGURE 9-3 Function—type multipart book order form

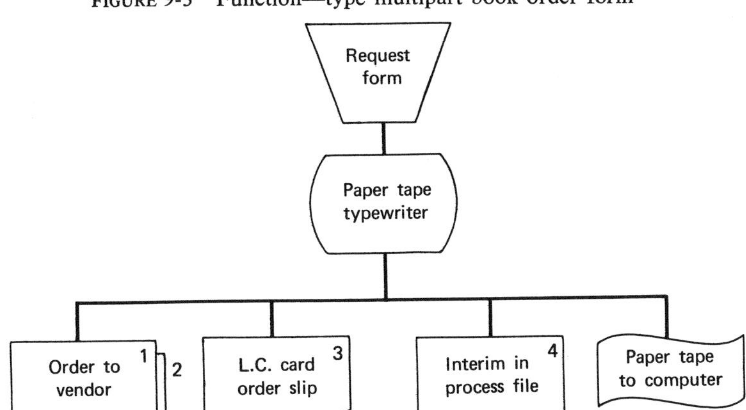

2. Maintains control over format of data to be entered. For example, if in a certain area it is required that four characters of information be inserted into the record and if the operator only enters three characters of information, the machine will stop and not continue until the field has been filled.

3. The control tape automatically inserts into the input tape the necessary control characters to identify the beginning and end of each variable-length field.

Satisfaction of the requirements resulted first in a multipart book order form (Figure 9-2) that was designed to fulfill the existing requirements as well as to provide the necessary input for computer processing. Figure 9-3 diagrams the function of typing the book-order set and the disposition of the output in terms of external and internal requirements. Parts one, two, and three of the multipart form may be considered as satisfying external requirements: copy 1 to be used for the vendor's records; copy 2 to go to the vendor for return with the book or for use as a report on the book's availability; and copy 3 to be used for ordering Library of Congress catalog cards and on return, as notification to the individual requestor of the book's availability. Part 4 is filed in the interim in-process file until the order appears on the weekly computer-produced in-process list.

The paper tape provides the input for various other requirements satisfied by computer-produced reports.

1. *Current On-Order File.* Under manual procedures continuous control and updating of such a file is nearly impossible. However, it is possible to update the file and keep it current with the computer-based system. Further, the computer-produced on-order file reduces the manual

FIGURE 9-4 Inprocess list

TITLE / ORDER NUMBER-DEPTREQ-AUTHOR	DATORD	DATAVL	DCLMNO	DATREC	DATCAT
ACETYLENIC COMPOUNDS 6807873 RUTLEDGE, THOMAS F.	120668			031969	
ACHIEVEMENT IN AMERICAN SOCIETY 68W0524 ROSEN, BERNARD C.	082068		0318 1		
THE ACHIEVEMENT OF GREECE, A CHAPTER IN HUMAN EXPERIENCE 68W0683 GREENE, WILLIAM CHASE	091168			092568	
ACOUSTIC PROPERTIES OF WAKES (PHYS. OF SOUND IN THE SEA PART 4) 6807696 WILDT, R.	112668		0318 1		
ACOUSTICAL HOLOGRAPHY. VOL. 1 6902638 METHERELL, A. F.	040269				
ACTES DE LA 3E CONFERENCE INTERNATIONALE DE RECHERCHE OPERATIONELLE 6802121 INTERNATIONALE CONFERENCE DE RECHERCHE OPERATIONELLE, 3RD, OSLO,	032168		0318 1		
ACTINOMYCIN (NATURE FORMATION AND ACTIVITY) 6901457 WAKSMAN, SELNAN A.	030469			031969	
ADAM'S ANCESTORS 6901462 LEAKEY, LOUIS SEYMOUR BAZETT	030469			040969	
ADHESION (INST. OF MECH. ENG. V.178, PT.3F) 6902746 CONVENTION ON ADHESION. LONDON 1963	040769				
ADMINISTERING INSTRUCTIONAL MEDIA PROGRAMS 6903080 ERICKSON, CARLTON W. H.	041869				
ADMINISTRATIVE FINANCIAL MANAGEMENT 6900302 BRADLEY, JOSEPH	011669			021869	
ADMINISTRATIVE SYSTEMS MANAGEMENT 6902812 PEMBERTON, LEROY A.	040769			050669	
ADVAITA AND VISISTADVAITA 6800324 SRINIVASACHARI, S. M.	011568		0318 1		
ADVANCED ACCOUNTING PROBLEMS 6903304 CHAYKIN, IRVING J.	042969				
ADVANCED ELECTRICITY AND MAGNETISM (36056) 6901983 DUFFIN, W. J.	031969				
ADVANCED PHYSICAL CHEMISTRY A SURVEY OF MODERN THEORETICAL PRINCIPLES 6902228 BLINDER, S. M.	032469	060069			
ADVANCED PROPULSION SYSTEMS FOR SPACE APPLICATIONS (AIAA LOS ANGELES SECTION MONO.) 6902426 KNAPP, DAVID EDWIN	033169	TL 78275	.K55	042869	51369
ADVANCED QUANTUM MECHANICS AND PARTICLE PHYSICS. V. 2 AND STANDING ORDER FOR FUTURE VOLUME 6802361 EISELE, JOHN A.	032868		0318 1		
ADVANCED SPACE EXPERIMENTS (ADVANCES IN THE ASTRONAUTICAL SCIENCES. V. 25) 6903024	041669	TL 787	.A6 A2 V.25	041669	51369

FIGURE 9–5 Claim letter

RENSSELAER POLYTECHNIC INSTITUTE
LIBRARY BOOK ORDER DEPARTMENT
TROY, NEW YORK 12181, U.S.A.
(518 270-6427)

MAY 23, 1969

GORDON & BREACH PUBLISHERS, INC.
150 FIFTH AVENUE
NEW YORK
N. Y. 10011

GENTLEMEN-

THIS IS THE SECOND REQUEST BY THE RENSSELAER POLYTECHNIC INSTITUTE LIBRARY BOOK ORDER DEPARTMENT AS TO THE STATUS OF THE FOLLOWING ORDER.

ORDER NO.- 6807661

AUTHOR - LIVSHITS, M. L.

TITLE - THE EFFECT OF IONIZING IRRADIATION UPON THE FUNCTIONS OF THE CENTRAL NERVOUS SYSTEM

PLEASE ADVISE US IMMEDIATELY AS TO THE STATUS OF THIS ORDER.

( ) NOT OUR PUBLICATION, ( ) CANCELLED,

( ) OUT OF STOCK, HOLDING ORDER, ( ) OUT OF PRINT,

(✓) NOT YET PUBLISHED, ( ) OTHER-

THANK YOU.

SINCERELY YOURS,

(MRS.) LOUISE WEIMER
ACQUISITIONS LIBRARIAN

FIGURE 9-6 Fiscal report

UPDATE RUN 017	ORIGINAL ALLOCAT'N	FUNDS ADDED YEAR TO DATE	FUNDS WITHDRAWN YEAR TO DATE	TOTAL ALLOCAT'N	PREVIOUS BALANCE 10/15/68	FUNDS ADDED THIS MONTH	FUNDS WITHDRAWN THIS MONTH	EXPENDITURES YEAR TO DATE MAIN	EXPENDITURES YEAR TO DATE DEPT	ENCUMBERED MAIN	ENCUMBERED DEPT	CURRENT BALANCE 10/28/68
185.180	-156.41	147.66		-8.75	-142.66	147.00						5.00
185.390	3276.89	17000.00	63.65	20213.24	16561.41	17000.00	62.00	2393.68		3538.72		14280.84
185.580	754.47	33.28		787.75	-33.28	33.00		787.75				
190.410	27500.00		817.13	26682.87	20930.42		815.00	2148.65		4509.58		20024.64
760.090	9220.76		4591.24	4629.52	4334.61		4588.00	299.67		532.55		3797.30
761.010	821.90		749.90	72.00	14.38		749.00	52.12		5.50		14.38
790.014	628.82		12.76	616.06	156.78		12.00	327.32		352.96		-64.22
790.064	1004.43		51.96	952.47	635.48		51.00	14.30		353.45		584.72
790.084	6483.07		30.30	6452.77	4772.82		30.00	1074.49		611.70		4766.58
790.094	34.91		1.44	33.47	26.70		1.00	8.21				25.26
790.134	189.59		10.53	179.06	177.06		10.00			3.00		176.06
790.170	602.72	1000.00	235.84	1366.88	1386.47	1000.00	235.00	150.15		66.10		1150.63
790.190	2130.30	597.80	15.45	2712.65	2317.36	597.00	15.00	113.94		281.35		2317.36
790.200	4576.36		71.09	4505.27	1753.64		71.00	1348.50		1418.09		1738.68
790.214	177.62			177.62	91.22					86.40		91.22
790.240	5000.00			5000.00	5000.00							5000.00
790.250	13160.00		3198.91	9961.09	2881.83		3198.00	5433.05		1653.91		2874.13
790.260	11214.43		7362.42	3852.01	-519.68		7361.00	701.31		3237.24		-86.54

END OF ACCOUNTING PERIOD (MONTH OR FISCAL YEAR)

CUMULATIVE FISCAL YEAR TO DATE EXPENDITURES FOR CONTINUATIONS UNDER 190.410 TOTAL 1207.07

filing operation to the maintainance of a weekly interim in-process file. The computer-produced in-process list provides a biweekly cumulated list of orders including any changes made after the original typing of the order. Figure 9-4 is a sample page from this listing.

2. *Claiming Procedure (Order Follow-up).* Because the conditions under which a claim was initiated had to be specified in detail, this requirement could be well satisfied by a computer-based record (see Figure 9-5).

3. *Fiscal Reports.* In order to keep an accurate and current account of the encumbrances and expenditures of the acquisitions system, a variety of fiscal reports were required. As the record-keeping function increased in volume and complexity the more amenable it became to computer-based manipulation. One of the fiscal reports includes statements of funds added, withdrawn, expenditures year to date, encumbered funds, and previous and current balances for each account (See Figure 9-6).

4. *Operating Statistics.* As the volume of book orders increased the task of keeping statistics obviously became more complex. The requirement of maintaining a variety of routine statistical reports and producing special reports as needed was suited to a computer control (An example of a statistical report is supplied by Figure 9-7).

COMPUTER PROCESSING, STORAGE AND OUTPUT MEDIA

It was established that the initial computer input was to be the punched-paper tape produced by the tape typewriter and that the records were then to be stored on magnetic tape for processing by the computer. The design of the machine-readable record and the system for processing it had to be developed within the framework of these two parameters, paper tape for input and magnetic tape for processing and storage.

In designing the machine record and specifying the processing the first step was translation of the record from the punched-paper tape into the magnetic-tape code, which is a character-by-character translation. Here advantage was taken of the computer's abilities to structure in a predetermined format, a magnetic-tape record that makes provision for all essential data elements whether or not they are present in the record being translated. If the translated magnetic-tape record was to be readily updated throughout, if each new record could contain any or all of the allowable data elements, and if records could be entered in random order, additional parameters had to be considered to further develop the system.

As a result, two major programs were developed: the translation program, including formatting and editing of the record, and a general process-

FIGURE 9-7 Operating statistics report

5/28/69 UPDATE RUN 30

DEPARTMENT	BOOKS ORDERED MAIN	DEPT	BOOKS RECEIVED MAIN	DEPT
			9	
ARCH				
ACHV				
BE	33	1	27	1
BIOL	505		429	
CHEM	303	4	258	3
CONT	417	11	416	11
ECON	144		146	
ELEC	32		27	
ENG			10	
FCT	114	23	82	18
GEN	1611		1419	
GEOL	453	22	374	23
HIST	583		542	
HSS			25	
L&L	920		733	
MS	158	4	82	5
MGT	1078		837	
MAT	65		46	
MATH	576	2	482	2
MECH	127		127	
MUS				
NES	37	41	22	34
PHIL	195		198	
PHYS	356	71	275	64
PSY	24		26	
REF			1	
SCI			21	
SCIT	2	54	4	32
SE	123	19	111	[illegible]

ing program. The two programs had to be separated by another program because of two other seemingly incompatible requirements: that records be readily updated and that they be accepted in random order. To update a record it first had to be found in the file. Finding it required that it be uniquely identifiable and implied, with sequential access storage (magnetic tape), that the records were in a specified order. Thus each record had to have a code (handle) by which it could be located in the computer file. After a record was found, the specific part of it that was to be updated had to be identifiable and accessible within the record itself.

Computer records come in two basic formats: fixed length and variable length. Using the former to satisfy the updating requirement was the logical choice since most of the records would contain predictable elements of data; for example, author, title, and publisher. A size limit was set that was large enough to accommodate the longest expected variable of each element. Each element then had its own field (fixed) in the same place in every record regardless of whether the field contained valid data or merely blanks supplied during translation. From a programming standpoint, in this application, fixed-length fields and records were considered much more convenient to work with than variable-length fields and records.

At this stage in the design of the machine record and the processing system the following points were decided upon:

1. Each record to have a unique number and was to contain fixed-length data fields.

2. A computer translation program that would accept paper-tape input developed to (a) translate the punched code into computer code, (b) edit the translated data for validity, and (c) format the data onto a magnetic tape for further processing.

3. A sort program to arrange the records by their unique numbers.

4. A processing program to update the records.

5. Additional programs to produce from the update master file output needed to satisfy the overall requirements of the system; for example, a printed in-process list and a current accessions list.

To explain more fully the computer translation program (item 2), editing the data elements was possible with the parameters established for valid data in any given field. If a price field, for example, was defined as one containing only digits or the letters GIFT, then a gift book coded as FREE in that field would be considered an error by the edit feature of the program. Such invalid data were to be printed out in an error listing for correcting the input.

FORMATTING RECORDS

The formatting of the records was done by the programmer after the data elements had been established for the acquisitions system. It was here that the programmer played an important role. Balancing the needs of the system against the capabilities of the computer, he developed a tape format that would take maximum advantage of the machine's size, speed, and logic features to meet the requirements of the system. Figure 9-8 is a schematic rendering of what such a record looks like.

All but three of the fields in the figure are supplied directly by the paper-tape input. Content in the three fields named CONNO (control number), CLAIMS, and STATUS, each indicated by arrows, is supplied entirely under computer program control as a product of data processing but there is a difference between the data in CLAIMS and that in either CONNO or STATUS. The data in CLAIMS is *informative;* that is, if the record were printed out, the characters in the field would supply information directly to the reader without need of translation. It should be remembered that one of the original demands made of the record was that it contain all the codes or *indicators* enabling it to be processed. CONNO and STATUS are examples of indicator fields. One feature of the system/360 computer that makes it a powerful data processing tool is its ability to access, test, and manipulate individual *bits* within a *byte.* A byte is composed of eight bits and represents one alphabetic or special character, two decimal digits or eight binary bits of information. A byte for the letter A, for example, has the bits arranged thus: 11000001. By assigning each bit position a 1 or a 0 (on, off), eight separate conditions or any combination thereof may be carried in one byte. The two fields STATUS and CONNO are each two-byte-long "message centers" with a combined potential for identifying 32 separate conditions. The data in STATUS and CONNO, then, are indicative rather than informative. If a particular record were printed out, its bit patterns in STATUS and CONNO might or might not have a graphic representation on the printout and even if they did, the graphic would mean nothing in itself until it could be reduced to its own bit pattern and then interpreted. The combined four bytes of CONNO and STATUS may be used to indicate internally a record's condition and stage of processing throughout the time it is under program control. Most conditions are best shown by the use of the CONNO and STATUS bit pattern "flags" rather than by relying on the presence or absence of information within the record. The presence of price data, for example, does not show that an item has been paid for, but a bit that can be set only when the invoice is cleared and paid shows the item as paid. The following table shows

FIGURE 9-8 Machine-readable record format—diagram of field positions in translated record

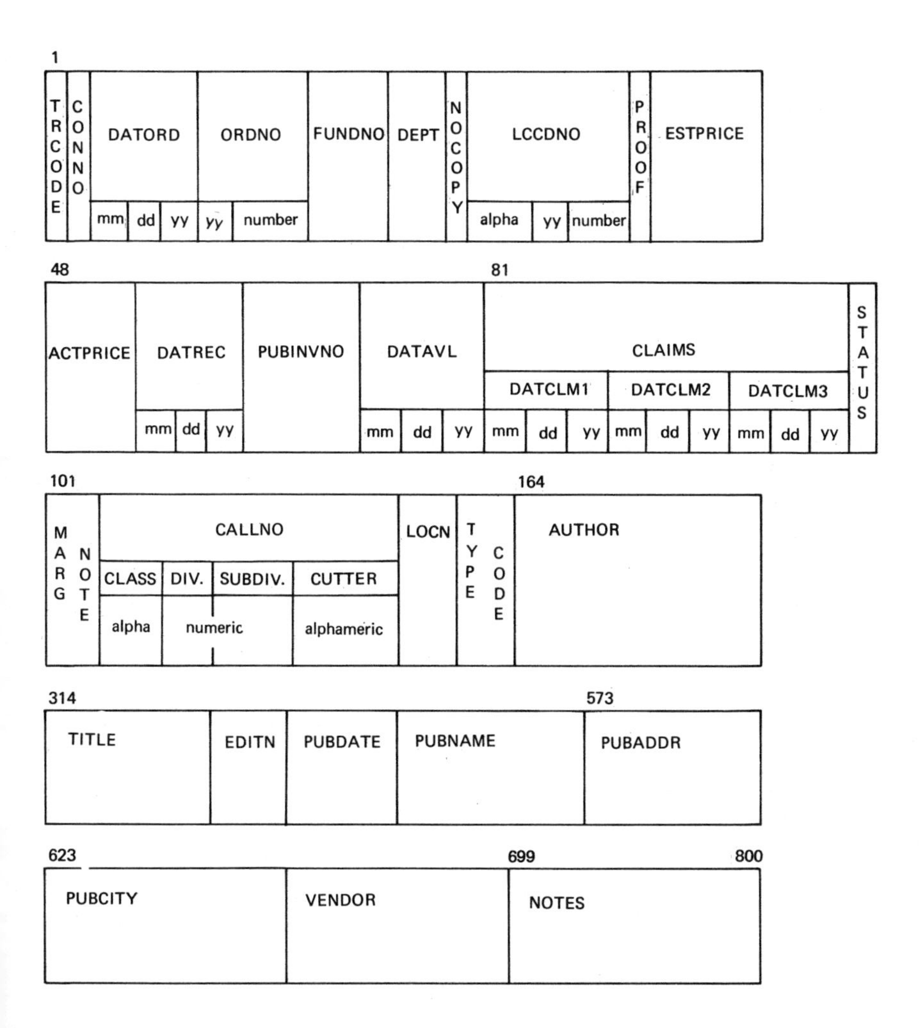

some of the conditions that the bits in CONNO and STATUS are used to indicate.

Bit Flags in CONNO and STATUS

	Bit	Condition
CONNO	0	On order
	1	Claimed
	2	Received—in process
	3	Invoice cleared and paid
	4	Cataloged
	5–15	Unassigned
STATUS	0	Departmental copy
	1	Reserve request
	2	General library copy
	3	Gift
	4	Gift plate
	5–15	Unassigned

SORTING AND UTILITY PROGRAMS AND MACRO INSTRUCTIONS

The process of sorting is accomplished by utility programs. These programs usually are supplied by the computer manufacturer. They are very general and flexible with broad computer applications such as sorting a file, merging files, and punching cards. There are usually few limitations on the use of such programs and normally all the user needs to do is to describe the characteristics of his file and individual record to a particular utility program. In using a utility program for a sorting operation the input and output devices are designated first, the size of the file is given, then the individual record is specified as to its length, the position and length of the fields to be sorted (given in the order in which they are to be printed out—such as by order number, author, and title); and the sequence of the sort whether in ascending or descending order.

Related to the utility programs are the macro instructions. A definition of macro, "of or involving large quantities," implies its use. In computer terminology a macro is an instruction having the capability of generating more than one machine-language instruction and is incorporated under a coded name into the monitor system of the particular machine. Thereafter when the macro's code name command is given, the monitor system expands it into the predefined set of instructions. Some commonly used macros are open, close, read, write, get, put, print, and punch. Generally in an actual instruction these macros would be followed by one or more file names showing what device or file the specified command is to act on.

The macro facility in general provides convenient shortcuts for the programmer and wherever necessary he can write his own macros. If, for example, under certain conditions *A* were to be replaced by *a* many times in several different programs, the programmer could elect to write a macro called "replace" which would contain all the instructions necessary to replace *A* with *a*.

For example, a utility sort program specifying publisher and title fields within the acquisitions transactions tape could arrange the file by publisher and for each publisher the titles on order to generate a list of in-process orders.

The sort procedure operates with a nondestructive readout. At the conclusion of a sort there is not just a single file in a different order from that with which the sort began but two files. The original file is not erased or in any way destroyed in the process of being sorted. In the acquisitions system a utility sort is used prior to the update program to produce a file sorted first by order number and then within the order number by the transaction codes to bring each record and its associated updating information together in a predetermined order for processing by the update program. As previously stated the system was to accept updating information in random order but a sort was necessary between the translation and update programs because the records on both the master and transactions files are in sequential order to expedite matching of update data with the proper master record.

UPDATE AND UPDATE PROGRAM

The term "update" is very general and may be roughly equated to the expression "file maintenance." In a strict interpretation this involves modifying a master file with current information according to a specified procedure that could include an exchange of characters for blanks or if the master record is being corrected, an exchange of valid characters for invalid ones. Using presorted input the steps of a file maintenance program are to match each transaction to the proper master record, identify the type of transaction (price or date received), and enter new data or corrected data.

The designation "update program" was given the file maintenance procedure of updating combined with other procedures. While the primary achievement of the update program is a current on-order file, such features as bit testing and manipulation and speed of processing make it practical to satisfy other system requirements concurrently. Output is produced as a by-product of processing each record including such useful procedures as statistics on books ordered and received, reports on the amounts encumbered and spent from various funds, lists of errors in input data, and a file of items for which processing has been completed.

CHANGES IN THE ACQUISITIONS SYSTEM

From the design of the acquisitions system under discussion a core of programmed processing emerged that included facets of the operations in the acquisitions subsystems. In a computer-based system the core of program processing will often cut across the boundaries of subsystems as they existed before the new system was initiated. In this acquisitions system the various operations and functions of the manually maintained subsystems had to be realigned with the revised subsystems for the newly designed system to be effective. As a case in point we can consider the four acquisitions subsystems of preorder searching, ordering, accounting and reporting, and receiving and checking. The changes within the redesigned system are shown as follows.

Before	*After*
Preorder searching	Unchanged
Ordering	Ordering, I/O preparation and processing
Accounting and reporting	Data processing
Receiving and checking	Receiving, I/O preparation and processing

At least one major operation of ordering, the maintenance of an in-process file, has been taken over by the new subsystem data processing. Similarly the operation of maintaining account balances, previously maintained by accounting and reporting is now absorbed by data processing. The original requirements of the system are still being satisfied and further the new system has the potential of satisfying increases in these requirements.

ASPECTS OF INTEGRATED SYSTEMS DESIGN

In planning the design of the acquisitions system that we have been discussing, the total-systems approach was used in considering the placing of the routine and data processing operations of the system under machine control.

As part of the design of the computer-based acquisitions system

FIGURE 9-9 Master machine-readable circulation card

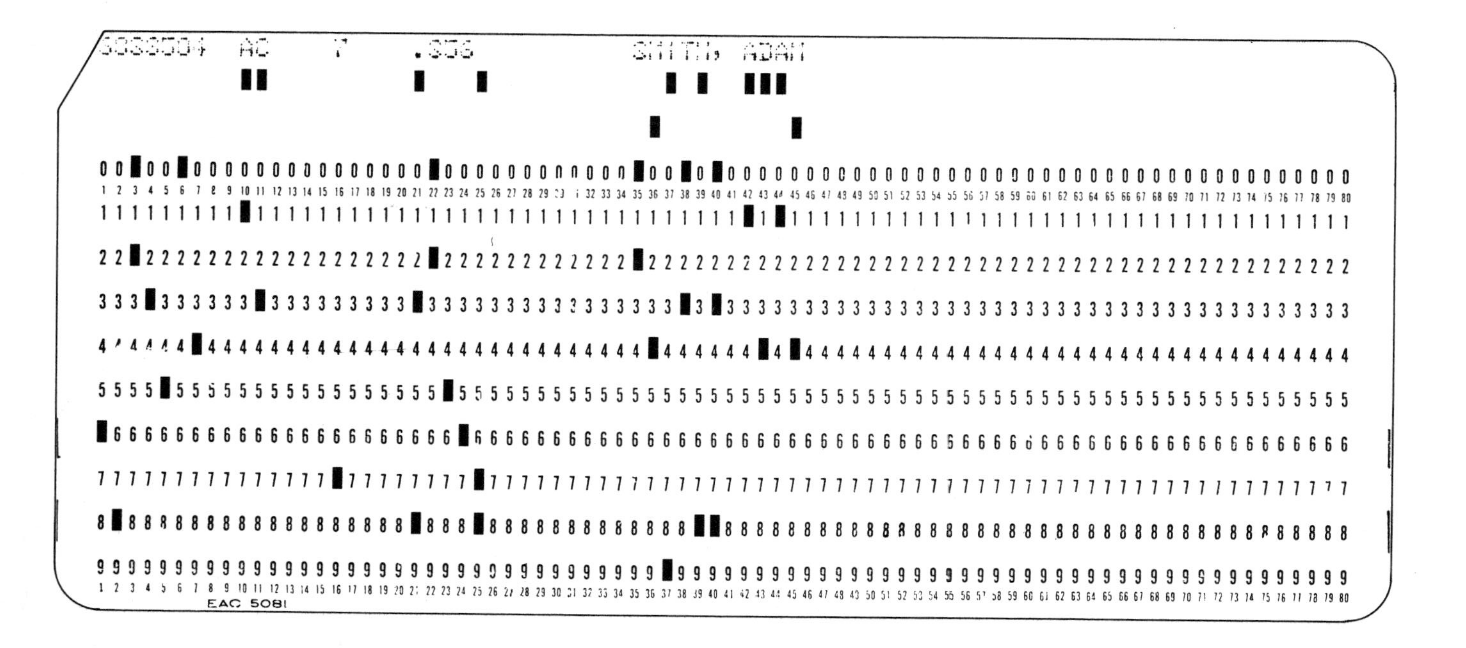

it was decided that punched cards (Figure 9-9) for the circulation system should be produced from the magnetic-tape acquisitions record file after it had been updated with call numbers. The program produced a master machine-readable circulation card for a currently processed book. This card, which lists the accession number, the author or title, and the call number, was inserted into the book. This procedure was the first phase in implementing the computer-based circulation system while still operating under manual conditions. Because, new books are expected to circulate more than the old[2] and because most new and circulating materials are equipped with punched cards, the automated circulation system should be fully operational at the time of installation.

Another output produced after the machine-readable acquisitions record has been updated with call numbers is the accessions list (Figure 9-10). It is arranged by call number and provides the author and title of each book. This bimonthly listing is distributed by the reference librarians as a current awareness service to users. It is different from the in-process list (Figure 9-4) in that the accessions list is a shelf list of newly cataloged books. It is intended to allow persons interested in a given subject to scan a pertinent block of call numbers and be aware of new library acquisitions that may relate to his field of interest.

As a requirement of the machine-readable acquisitions file it was decided that the file should be a historical (retrospective) record primarily for the possible use of these records as the means to find and "lift" full machine-readable cataloging data from MARC tapes or from some other library's cataloging data bank. This is seen as a way to save time and effort when the decision is made to convert existing cataloging records to machine-readable form.

[2] Richard W. Trueswell, Two Characteristics of Circulation . . . , *College & Research Libraries*, **25**:287, (July, 1964).

FIGURE 9-10 Accessions list

```
                         RENSSELAER LIBRARIES
      NEW BOOKS ADDED TO COLLECTIONS DURING DEC. - JANUARY PAGE     3

B  3216 .C33 P513 V.1,3
    CASSIRER, ERNST
         PHILOSOPHY OF SYMBOLIC FORMS. 3 VOLS.
B  3245 .F24 A75
    IGNACIO, ANGELELLI
         STUDIES ON GOTTLOB FREGE AND TRADITIONAL  PHILOSOPHY
B  3376 .W564 M6
    MORICK, HAROLD
         WITTGENSTEIN AND THE PROBLEM OF OTHER MINDS
B  3614 .C72 E57
    CROCE, BENEDETTO
         PHILOSOPHY, POETRY, HISTORYAN ANTHOLOGY OF ESSAYS
B  4568 .U54 I4
    ILIE, PAUL
         UNAMUNO  AN EXISTENTIAL VIEW OF SELF AND SOCIETY
B  5131 .N3 1964
    NARAVANE, V. S.
         MODERN INDIAN THOUGHT
B  8233 .L6
    LOBKOWICZ, NIKOLAUS
         THEORY AND PRACTICE  HISTORY OF A CONCEPT FROM ARISTOTLE TO
BC    6 .F67
    FOSTER, MARGUERITE H.
         PROBABILITY, CONFIRMATION AND SIMPLICITY
BC   71 .B3
    BEARDSLEY, MONROE C.
         PRACTICAL LOGIC
BC  135 .G4
    GENTZEN, GERHARD
         INVESTIGATIONS INTO LOGICAL DEDUCTIONS
BC  135 .W37 1960
    WEINBERG, JULIUS RUDOLPH
         AN EXAMINATION OF LOGICAL POSITIVISM
BC  177 .H6
    HODWETT, EDWARD
         THE ART OF PROBLEM SOLVING
BD  161 .C48
    CHISHOLM, RODERICK M.
         THEORY OF KNOWLEDGE
BD  220 .S2
    SCHEFFLER, ISRAEL
         SCIENCE AND SUBJECTIVITY
BD  638 .G7713
    GUITTON, JEAN
         MAN IN TIME
BF   20 .A78 1966
    SANDERS, ANDRIES FRANS
         ATTENTION AND PERFORMANCE
BF  149 .S3713
    SERVADIO, EMILIO
         PSYCHOLOGY TODAY
BF  203 .W4
    WEETHEIMER, MAX
         DREE ABLAANDLUNGEN ZUR GESTALTHEORIE ERLANGEN
```

chapter ten

SYSTEMS DESIGN—COMPUTER-BASED SERIALS SYSTEM

Designing a computer-based serials system follows the prescribed pattern: the establishment of requirements, the provision of forms and the determination of the manual and machine procedures that will satisfy the requirements.

THE REQUIREMENTS OF THE SERIALS SYSTEM

The primary requirement of the serials system is to provide users with an accurate and current file of the library's serials holdings including all the necessary bibliographic information. In the instance we shall discuss, the file will be a union catalog containing the data required on all serials in both the general library and the branch libraries. The data in this file should be available not only in the general library but also in branches and to other possible users.

Second, the system should be able to provide the means whereby the user as well as the library staff may request and receive within a reasonable time special catalogs or reports of the serials holdings by such categories as call number, language, and subject; and it should also provide special listings by publishers, vendors, types of serials, status (such as active, inactive), and by the condition of receipt (gift, purchase, or exchange).

Third, the system should be able to guarantee that all issues not received by the library will be claimed promptly and on a set schedule and also that the renewal of all serial subscriptions will be controlled by the system.

Fourth, the system should be able to predict the time of receipt of a journal issue with accuracy thereby establishing a positive control over the timely receipt of all issues.

Fifth, the system must be able to maintain and control all the accounting records necessary for its own internal use and be able to report to the comptroller the information required for payment of invoices.

Sixth, the system should be able to provide the binding department with the necessary information and records for those journals that are to be bound. It also should be able to provide such information and records on a scheduled basis in order to maintain an even workload for the binding staff.

Seventh, the serials system should have the ability to handle the many changes that occur in the various serials titles.

COMPILING SERIALS RECORDS— INPUT TO THE SYSTEM

With the establishment of the requirements of the system the next task is to determine what information is necessary as input to the computer system if it is to satisfy the requirements as outlined. It should be determined where this information can be obtained if existing records are inadequate and what format and under what restrictions the input of information should be placed. The input worksheet should be designed to obtain the maximum amount of information with the least cost. The form finally adopted should contain all the information necessary for the functioning of the system.

INPUT WORKSHEET

Figure 10-1, Serial Control Record Worksheet, provides for all the information as input to the computer system. The following is a description of the procedures for compiling this information, identification of the

FIGURE 10-1 Serial control record worksheet

DEWEY CALL NO. 020.5 L72

VERIFIED L. C. 1958-62 Vol. 77 E. P.

L. C. CALL NO. Z671.L7154

MAIN ENTRY
Library Resources and Technical Services

SHORT TITLE — LIB RESOURCES TECH SERV

HOLDINGS - START				HOLDINGS - END				PUBLICATION - START				PUBLICATION - END			
VOL.	ISSUE	MO.	YEAR	VOL.	ISSUE	MO.	YEAR	VOL.	ISSUE	MO.	YEAR	VOL.	ISSUE	MO.	YEAR
0001	0001	01	1957					0001	0001	01	1957				

LIB.	LOC.	LANG.	TYPE	STAT.	RET.	BIND	COLOR	ISSUE PHY. VOL.	INDEX	SUP.	INDEXED IN	FREQ.	NO. OF ISSUE VOL.	C.C.	J	F	M	A	M	J	J	A	S	O
00	1	02	30				0720	08	06		125 130	14	04		X		X		X		X			

REC.	GROUP NO.	DATE START	LAST PAYMENT	SUB. LGT.	REQ. DEPT.	FUND NO.	SUBS. PRICE
6	22						

ISSUING AGENCY
American Library Association
Publishing Dept.
30 E. Huron Street, Chicago 11, Illinois
SUPPLIER

RENSSELAER POLYTECHNIC INSTI

SERIAL CONTROL RECORD

A - ADDED F - FORMERLY ISSUED AS M - MERGED N - NOTES Z - SUPERSED
S - SUBJECT C - CONTINUED AS R - MEMBERSHIP X - SUPERSED

°S°Library Science

°A°American Library Association. Resources and Technical Services Division

°Z°Serial Slants and Journal of Cataloging and Classification

°R°American Library Association

bibliographic sources, and description of what each field of the record will contain.

COMPILING PROCEDURE

The original step in compiling the serials record starts when the compiler removes from the existing central serials file the holdings record, or the current check-in record, or both. The holdings card indicates any title changes or title variations. If changes have occurred, the compiler removes from the central serials file those cross-referenced records pertaining to the title variations or changes in question; that is, all of the library's records covering the history of this particular title are removed. Referring to the serials control record worksheet, the first box in the upper left-hand corner carries the Dewey call number obtained from the library's record. In the case of the call number, where there have been title changes or there are multiple subscriptions and the same call number has been used for each new change or subscription, it is necessary for the compiler to add to the call number a hyphen and an additional digit (-1, -2, -3, and so forth) to identify each title within the computer record. The next step in compilation is to verify the official title, if possible, in the *National Union Catalog* (NUC). If it is found there, it is used as the main entry for input to the computer. The LC call number, if found in NUC, is recorded in the upper right-hand corner of the form.

The compiler also indicates under the word "verified" the volume and year of NUC in which the title was verified. If the publication is not found in NUC, the next step is to search in the *Union List of Serials*. Here again, if the title is found, the entry is accepted as the main entry. If the entry is not found in the *Union List of Serials,* the search is continued in *New Serials Titles* and if necessary, in *Ulrich's International Periodicals Directory.*

The next item on the form is the short title which cannot exceed forty characters in length. The short title should conform to the title as it appears on the cover of the serial itself as an aid to checking in the issue. The abbreviations for periodical literature provided by the United States of American Standards Institute are used in establishing the short title.

Proceeding down the form the next blocks are Holdings-start and Holdings-end. This information is obtained from the library's holdings record. The information for the Publication-start and Publication-end blocks is obtained from NUC or the *Union List of Serials* and if not located, it is taken from other bibliographic sources.

The items in the next two horizontal lines of data are coded. The following defines the data elements and supplies examples of some of the code structure.

Keyword	*Description of Data Element*
LIB (Library)	Refers to the particular library or collection that maintains the file of a journal (00 = general library, 41 = business and economics).
LOC (Location)	The location in the library. This is used primarily to indicate where the journal is to be shelved (2 = current display rack). It is also used to identify the location to which it is routed previous to shelving (A = table of contents photocopy service).
LANG (Language)	The language or languages in which the journal is published (01 = Chinese, 20 = multilingual with English).
TYPE	Refers to form of publication—that is, abstracts, periodicals, transactions (10 = index or abstract, 40 = proceedings).
STAT (Status)	Refers to variations such as active, ceased publication, discontinued subscription, title change (1 = ceased publication).
RET (Retention)	Refers to the period of time of retention and conditions under which a journal is kept, that is, one year and discard, two years in branch and forward to general library, and similar variations in handling (3 = discard after three years).
BIND (Bindery)	Refers to the commercial binder used.
COLOR (Color)	The four-digit code for the color of binding cloth to be used.
ISSUE PHY VOL (Issues per physical volume)	Defines the number of issues that are bound in a physical volume.
INDEX	Refers to when and how the index is received and where it is found in the journal (0 = no index, 4 = separate, free on request).
INDEXED IN	Obtained from *Ulrich's International Periodicals Directory*. Here is indicated indexing services in which the journal is indexed or abstracted (175 = *Readers' Guide*, 095 = *Fuel Abstracts*).
FREQ (Frequency)	Defines the frequency of publication of each title—that is, monthly, weekly, quarterly, and so forth (01 = weekly, 16 = bimonthly).
NO. OF ISSUE VOL	The number of issues per bibliographic volume.

CC (Claim Control)	Indicates the waiting period at the end of which an issue is claimed (21 = one month late).
J F M A. . . (Jan., Feb., Mar., Apr.)	Used for forecasting the time of receipt of issues so that the computer can produce the check-in punched card in advance of receipt.
REC (Receipt)	The condition of receipt—for example, gift, purchase, membership, group purchase, exchange (4 = gift, 9 = group purchase).
GROUP NO.	If, for example, a journal is received as part of the package plan with the American Chemical Society, a unique two-digit number is assigned to the American Chemical Society and each title received in the package plan is assigned that number so that all journals received under the plan can be identified as a group for accounting purposes.
DATE START	Refers to the date that the last subscription started and controls the renewal of the subscription.
LAST PAYMENT	Used to verify that a title was renewed previously and when.
SUB LGT (Subscription length)	Used in conjunction with DATE START to determine next renewal date of a title.
REQ DEPT (Requesting department)	Used in connection with the FUND NO for accounting and statistical purposes. The LIB codes are used in identifying departments that have requested subscriptions either for housing in the departments or in the general library.
FUND NO. (Fund number)	Refers to the account against which the subscription is charged.
SUBS PRICE (Subscription price)	Refers to the cost of a subscription or last renewal.
ISSUING AGENCY	Full name and address of the issuing agency (publisher).
SUPPLIER	The name and address of the supplier or vendor when different from the issuing agency.

The balance of the serials control record worksheet is used for the additional bibliographic control data desired in this system. Entered in the last part of the form are the S-SUBJECT entries as found in NUC for each title. If the title is not found in NUC, the librarians revising the records must assign at least one compatible subject to each title. The A-ADDED entries used are those given in NUC excluding the added entry for editor.

Indication of title changes follow the subject and added entries. If there is a title change with the new title continuing old volume numbering, such title change is designated F-FORMERLY ISSUED AS, or

C-CONTINUED AS. On the other hand a title change with its own and new volume numbering is designated Z-SUPERSEDES or X-SUPERSEDED BY. If two titles are merged to form a new title, both titles would be entered under the code M-MERGED.

Any necessary notes that will facilitate the user's approach to a serial title are entered under the code N-NOTES.

After the information has been compiled on the serials control record worksheets, it must be verified and, if necessary, revised. No information is sent to the data processing operator until it has been reviewed and revised as necessary.

PROCEDURES AND EQUIPMENT IN GENERATING COMPUTER INPUT RECORD

After the data on the worksheet have been revised and all corrections made, it is forwarded to the data processing operator who prepares the input tape using the paper-tape typewriter. Reading from the serials control record worksheet the data processing operator types this information on a serials control record form simultaneously producing paper tape containing all of the compiled information.

There are three ways of correcting information entered into the paper tape. (a) If a character is incorrectly typed, the operator immediately may insert a delete character to erase the error and continue with the typing. (b) The operator at any time during the typing of a record may insert a special code which will indicate to the computer that this record is to be disregarded and that a substitute record will be entered. (c) At the end of the record the operator has the opportunity to correct any field that may have been typed in error by inserting the field control number followed by the correct information; a computer program will then insert the corrected information furnished by the operator, in the field indicated.

Each serials control record form produced concurrently with the paper tape is proofread for typographical errors at the end of the day's production. Any errors that are found during proofreading may be corrected by the operator at any time. The correction procedure is to punch five items into the paper tape; first, the code indicating to the program that this is a correction record; second, a code indicating the type of correction (A for an addition to a record, C for a correction to some field within the record, or D to delete some item of information from the record); third, the Dewey call number of the title to be corrected; fourth, the field control number of the field in error; and fifth, the corrected information. The correction record on paper tape will be used to make the necessary corrections, additions, or deletions to the machine-readable record identified by the Dewey call number.

DESIGN OF COMPUTER—SERIAL RECORD LOAD SYSTEM

The next phase of the serials automation system consists of designing the program required initially to load and format the serials records. Figure 10-2, Serial Record Load System, illustrates the generalized system.

Requirements of the system are (a) to produce a master file on magnetic tape containing all serials records loaded to date and arranged in Dewey call-number sequence; (b) to produce a report for verification of each serials record loaded in the current run; and (c) to produce a report of any record or records containing erroneous information (errors, duplicated records).

Inputs to the system are (a) a paper tape containing new serials records; and (b) A paper tape containing corrections, additions, and deletions to be made in existing computer records.

TRANSLATE AND EDIT PROGRAM OF THE SYSTEM

The first requirement of the program is to produce a report listing each error found in the editing procedure. Based on the editing report the data processing operator will correct the errors (as described previously) and will re-enter the corrections into the computer in a later run of the translate and edit program.

The second requirement is to assemble the input data into the proper format as developed for the master serials record. Figure 10-3, Master Serials Record, shows the layout of the computer record which consists of five sections. The first section of the record is fixed in length and contains the majority of the data characterized as "basic identification and control data." The second section is fixed in length and contains the accounting data including a five-year historical record of the payment of each serial subscription. (A five-year average of the increase in serials cost should permit forecasting of serials budgets with a high degree of reliability.) The third section of the master serials record, variable in length, contains bibliographical information on each title. This includes the official entry, all subject entries, selected added entries, title changes,

FIGURE 10-2 System chart—serial record load system

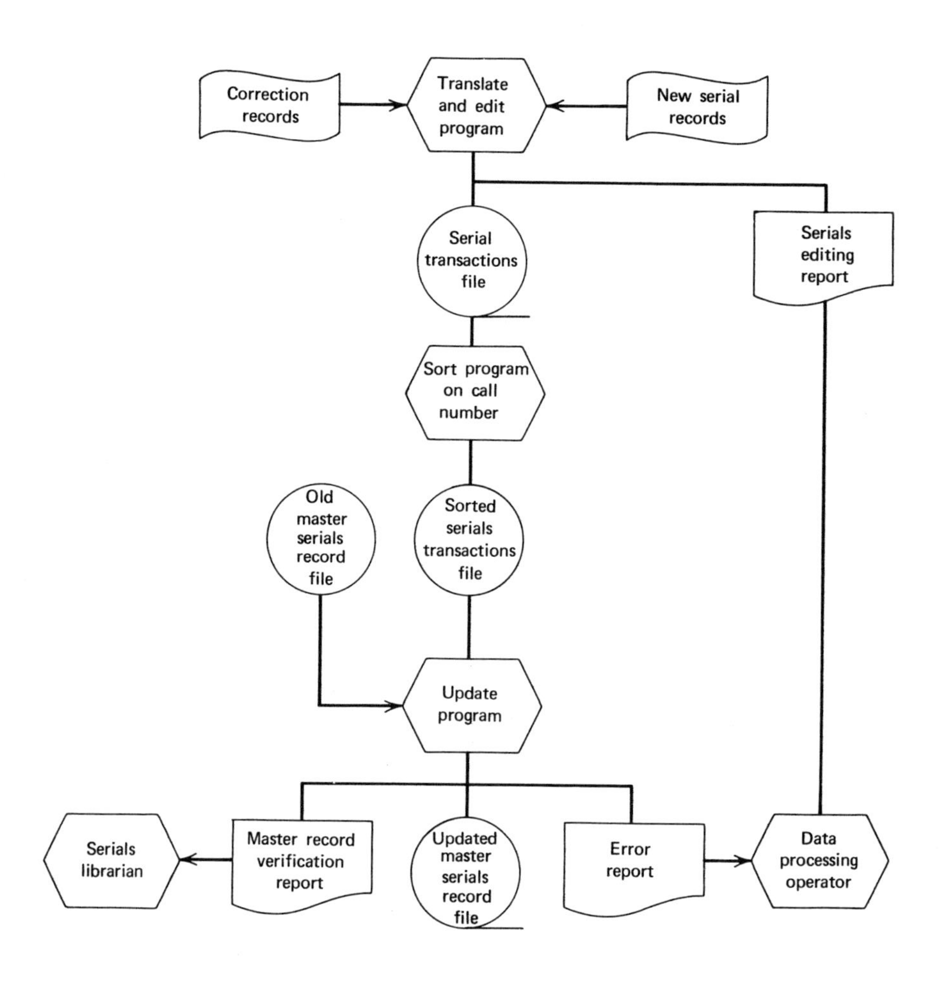

FIGURE 10-3 Master serials record

MASTER SERIALS RECORD

Dewey Call Number	L.C. Call Number	Short Title	Publication Dates								LIB	LOC	LANG	TYPE	STATUS	RET	Indexed In						
			Start				End																
			Vol	Iss	Mo	Yr	Vol	Iss	Mo	Yr													

Section 1, Basic Identification and Control Data

Cond. Rec.	Group No.	Date Subs. Start	Subs. Lgt.	Req. Dept.	Fund No.	Payment Record											
						Last Inv. No.	Date Last Pay	Subs Price	Date	Price	Date	Price	Date	Price	Date	Price

Section 2, Accounting Data

Main Entry	Subject Entries	Added Entries	Title Variations	Notes

Section 3, Bibliographical Information

Holdings								Freq.	Iss. Per Vol.	Claim Ctl.	Pub. Months	Date Last Rec.	Waiting Receipt and Claim Control						
Start				Current									Vol.	Issue	Mo.	Year	Claimed		
Vol	Iss	Mo	Year	Vol	Iss	Mo	Year				JFMAMJJASOND						Date	Date	Date

Section 4, Receipt and Claim Control

Co.	Color	Issues Phys. Vol.	Index	Sup.	Vol.	Year	Date Bind	INDEX	Month	Issue	Month	Issue

Section 5, Binding Control

and all notes. Each of these items is variable in length and each field, except the official entry field, may contain more than one entry. For example, there might be two or three subject entries for a particular title. The fourth section, variable in length, is the claiming and receipt section. This portion of the record is generated internally by the check-in program at the time it forecasts the receipt of each issue of a title. The fifth section, variable in length, contains the binding information, the greater portion of which is created by the check-in program at the time the updated information is received by the computer.

Inputs to the program are (a) paper tape of new serials records and (b) paper tape of corrections records.

Outputs of the program are (a) serials transactions file on magnetic tape containing the new titles that have been loaded during a given run plus corrections to previous titles that were loaded at an earlier date and (b) the editing report indicating what errors the computer has found during the edit process.

Functions of the program are (a) to read in the data from the paper-tape reader (since the codes that are produced by the paper-tape type-writer are not the same as the code structure used in the computer, the characters on this tape must be translated into the appropriate machine language); and (b) to edit each field of the serials record to make sure that it conforms to the requirements of length and content of the field and that the information as presented, is valid. For example, under the area LIB (library) it is possible that the operator typed a code that is not a valid departmental code—or in typing the MAIN ENTRY an invalid character was used or for an active serial the FREQ (frequency) code was omitted. Thus each item on the form, where it is possible, will be edited for validity and completeness.

SORT PROGRAM OF THE SYSTEM

The requirement of the program is to sort the new serial record file into Dewey call-number sequence.

Input to the program is the serials transactions file produced in the translate and edit program and containing new and corrected serials records.

Output of the program is a sorted serials transactions file containing new and corrected serials records in Dewey call-number sequence.

Function of the program is to sort the records contained in the magnetic-tape files into a desired order or sequence.

UPDATE PROGRAM OF THE SYSTEM

Requirements of the program are (a) to generate a new master serials record file; (b) to make corrections to existing records; and (c) to eliminate duplicate records.

Inputs to the program are (a) sorted serials transactions file of new and corrected serials records and (b) a sorted master file of existing serials records.

Outputs of the program are (a) a magnetic tape file containing all records, existing and new, (b) a printed listing of new and corrected titles loaded in this run, and (c) a printed listing of erroneous records.

Functions of the program are (a) to merge old and new records on the two input tapes into one output master file and (b) to correct any record in the old tape file prior to generating the new tape file.

DESIGN OF COMPUTER SYSTEMS—CATALOG REPORT GENERATOR SYSTEM

The catalog report generator system allows flexibility to the library staff and to the user by obtaining information from the system in the sequence and in the format wanted. This system is a medium by which users can exploit the informational potentialities of the serials control system.

Figure 10-4, Generalized Systems Chart of the Catalog Report Generator System, outlines the programs and their interrelationships required in the preparation of specified printed catalogs and reports.

Requirements of the system are (a) to provide the user with an accurate and current catalog of the serials holdings containing the necessary bibliographic information (see Figure 10-5, Serials Catalog) and (b) to provide the means whereby the user as well as staff members may request and receive special catalogs or reports of the serials holdings within a reasonable time (for examples, see Figure 10-6, Serials Listing by Subject; Figure 10-7, Serials Listing by Call No.; Figure 10-8 Serials Listing Departmental Libraries - Arch.).

Inputs to the system are (a) the master magnetic tape file of serial records and (b) the data control punched cards for selection of data and desired sequence and format of the report.

Outputs of the system are printed catalogs or reports. When printout of these catalogs or reports is required, it is only necessary for the serials librarian to indicate this by the appropriate code in a control card. The generator system based on this code will automatically select from the master record the data required for the printing of a given list.

FIGURE 10-4 Systems chart—catalog report generator system

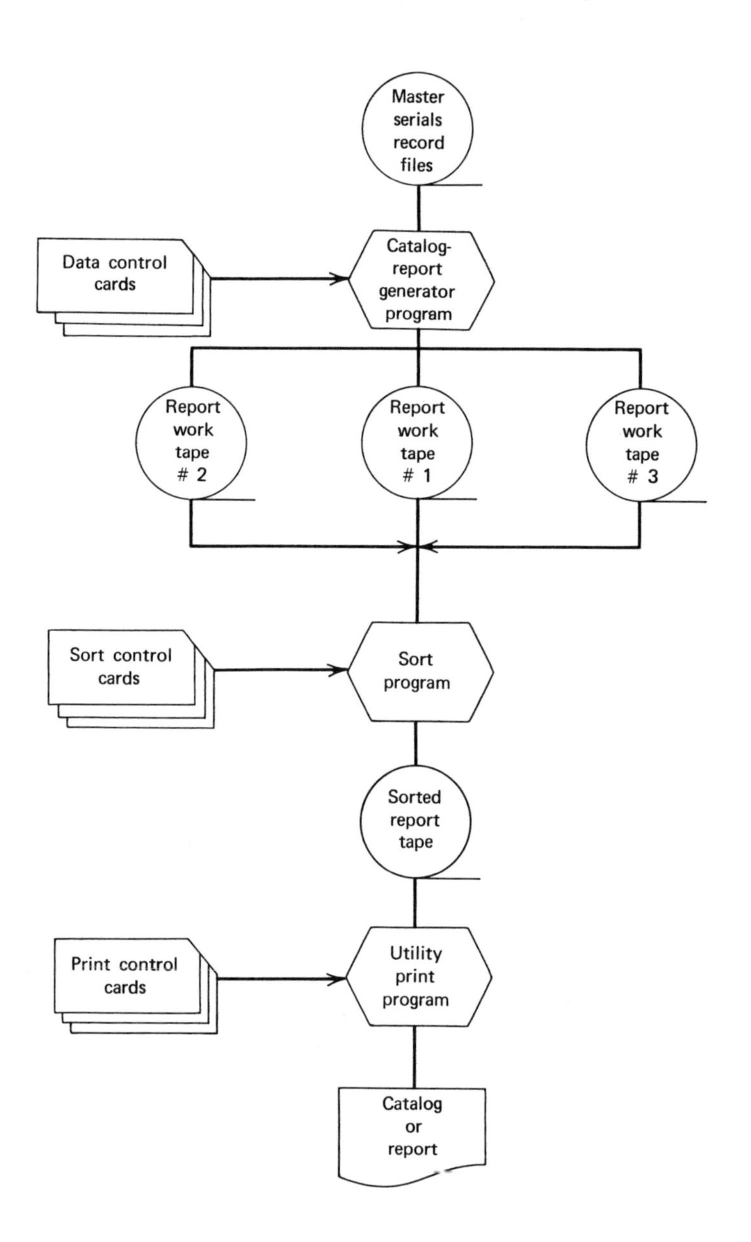

FIGURE 10-5 Serials catalog

SERIALS LISTING 05/13/69 PAGE 287

TITLE	DEPT	LOC	CALL NO.	HOLDINGS	CONTROL
NATURAL HISTORY LANG 22 PUBN DATES V1,1900-		DR	500.905 N36	V77,1968-	N111000
NATURE LANG 22 PUBN DATES V1,1869-		DR	505 N28	V5,1871-	N112000
	BIOL		505 N28	V209,1966-	
NATURWISSENSCHAFTEN LANG 04 PUBN DATES V1,1913-		DR	505 N26	V13,1925	N113000
NAVAL RESEARCH LOGISTICS QUARTERLY LANG 22 PUBN DATES V1,1954-			359.05 N318	V1,1954-	N114000
NAVY MANAGEMENT REVIEW LANG 22 PUBN DATES V1,1956-			353.7 D419N	V7N4,1962-	N117000
NEAR EAST REPORT LANG 22 PUBN DATES V1,1957-		DR	956.05 N35	V3,1959-	N118000
NEDERLANDSCH GEOLOGISCH MIJNBOUWKUNDIG GENOOTSCHAPE SEE GEOLOGIE EN MIJNBOUW					N119000
NEDERLANDSE CENTRALE ORGANISATIE VOOR TOEGEPAST-NATUURWETENSCHAPPELYK ONDERZOEK SEE APPLIED SCIENTIFIC RESEARCH APPLIED SCIENTIFIC RESEARCH. A. MECHANICS, HEAT, MATHEMATICAL METHODS APPLIED SCIENTIFIC RESEARCH. B. ELECTROPHYSICS, ACOUSTICS, OPTICS, MATHEMATICAL METHODS					N120000
NEMAN LANG 07 PUBN DATES V9,1960-			057 N43	V9,1960-	N121000
NEUES JAHRBUCH FUR GEOLOGIE UND PALAONTOLOGIE. ABHANDLUNGEN LANG 04 PUBN DATES V92,1950-	GEOL		550.53 N48	V127N2,1967-	N122000
NEUES JAHRBUCH FUR MINERALOGIE. ABHANDLUNGEN LANG 04 PUBN DATES 1950-		DR	549.05 N46	V108,1968-	N123000
NEUES JAHRBUCH FUR MINERALOGIE. MONATSHEFTE LANG 04 PUBN DATES N1,1950-		DR	549.05 N48	N1,1968-	N124000

FIGURE 10-6 Serials listing by subject

```
                                          03/11/69

AERONAUTICS--INDEXES
  AERONAUTICAL ENGINEERING INDEX
  AEROSPACE ENGINEERING INDEX

AERONAUTICS--SOCIETIES, ETC.
  AERONAUTICAL JOURNAL
  AERONAUTICAL SOCIETY OF GREAT BRITAIN. ANNUAL REPORT

AERONAUTICS--U.S.
  AERIAL AGE
  AERO DIGEST
  AMERICAN AVIATION

AERONAUTICS, COMMERCIAL
  AIRLIFT

AEROPLAND INDUSTRY AND TRADE--INDEXES
  AERONAUTICAL ENGINEERING INDEX

AEROPLANE INDUSTRY AND TRADE
  AERONAUTICAL ENGINEERING REVIEW

AESTHETICS
  BRITISH JOURNAL OF AESTHETICS

AIR CONDITIONING
  ASHRAE JOURNAL
  AIR CONDITIONING, HEATING AND VENTILATING
  AIR ENGINEERING

AIR WARFARE
  AIR UNIVERSITY QUARTERLY REVIEW
  AIR UNIVERSITY REVIEW

AIR--POLLUTION--ABSTRACTS
  APCA ABSTRACTS

AIR--PURIFICATION
  AIR ENGINEERING

ALMANACS, ENGLISH
  AN ALMANACK

ALUMINUM
  ALUMINIUM DEVELOPMENT ASSOCIATION. RESEARCH REPORT

ALUMINUM--ABSTRACTS
  ALUMINIUM LABORATORIES LIMITED, KINGSTON, ONT. ABSTRACT BULLETIN
```

FIGURE 10-7 Serials listing by call number

03/11/69

Call number	Title	
570.5 M33	BIOLOGICAL BULLETIN	DR
572.05 A512	AMERICAN ANTHROPOLOGIST	DR
574.05 B582	BIOMETRISCHE ZEITSCHRIFT	
574.05 B615	BIOCHEMICAL AND BIOPHYSICAL RESEARCH COMMUNICATIONS	DR
574.05 B71	BIOCHEMISTRY	DR
574.05 B93	BULLETIN OF MATHEMATICAL BIOPHYSICS	
574.083 B615	BIOMETRICS	
574.105 B615	BIOPHYSICAL JOURNAL	DR
574.1905 B613	BIOCHEMISTRY (BIOKHIMIIA)	
574.1905 B615	BIOCHIMICA ET BIOPHYSICA ACTA	DR
574.1905 B616	BIORHEOLOGY	DR
574.19105 B56	BIOPHYSICS	
574.19205 B56	BIOTECHNOLOGY AND BIOENGINEERING	DR
574.205 B752	BRITISH JOURNAL OF EXPERIMENTAL PATHOLOGY	DR
575.105 B56	BIOCHEMICAL GENETICS	DR
580.72 B789C	BOYCE THOMPSON INSTITUTE FOR PLANT RESEARCH. CONTRIBUTIONS	
605 B335	BATTELLE LIBRARY REVIEW	STR
605 B336	BATTELLE TECHNICAL REVIEW	DR
610.5 A51	AMERICAN JOURNAL OF TROPICAL MEDICINE AND HYGIENE	DR
612.015 B542	BIOCHEMICAL MEDICINE	DR
612.01505 B615	BIOCHEMICAL JOURNAL	DR
612.05 A512	AMERICAN JOURNAL OF PHYSIOLOGY	DR
612.05 B62	BIOCHEMICAL BULLETIN	STR
614.05 A51	AMERICAN JOURNAL OF PUBLIC HEALTH	STR
614.05 A51	AMERICAN JOURNAL OF PUBLIC HEALTH AND THE NATION'S HEALTH	
616.05 B13	BACTERIOLOGICAL REVIEWS	DR
620.105 S933	ACADEMIA REPUBLICII POPULARE, ROMINE. INSTITUTUL DE MECANICA APLICATA.	
620.11 AM5B	ASTM BULLETIN	
620.112205 B862	BRITISH CORROSION JOURNAL	DR
620.5 A188AM	ACTA POLYTECHNICA. APPLIED MATHEMATICS AND COMPUTING MACHINERY SERIES	
620.5 A188CB	ACTA POLYTECHNICA. DIVIL ENGINEERING AND BUILDING CONSTRUCTION SERIES	
620.5 A188CH	ACTA POLYTECHNICA. CHEMISTRY INCLUDING METALLURGY SERIES	
620.5 A188CH	ACTA POLYTECHNICA SCANDINAVICA. CHEMISTRY INCLUDING METALLURGY SERIES	
620.5 A188CI	ACTA POLYTECHNICA SCANDINAVICA. CIVIL ENGINEERING AND BUILDING	
620.5 A188EE	ACTA POLYTECHNICA. ELECTRICAL ENGINEERING SERIES	
620.5 A188EL	ACTA POLYTECHNICA SCANDINAVICA. ELECTRICAL ENGINEERING SERIES	
620.5 A188MA	ACTA POLYTECHNICA SCANDINAVICA. MATHEMATICS AND COMPUTING MACHINERY	
620.5 A188ME	ACTA POLYTECHNICA. MECHANICAL ENGINEERING SERIES	
620.5 A188ME	ACTA POLYTECHNICA SCANDINAVICA. MECHANICAL ENGINEERING SERIES	
620.5 A188PA	ACTA POLYTECHNICA. PHYSICS AND APPLIED MATHEMATICS SERIES	
620.5 A188PH	ACTA POLYTECHNICA SCANDINAVICA. PHYSICS INCLUDING NUCLEONICS SERIES	
620.5 A188PN	ACTA POLYTECHNICA. PHYSICS INCLUDING NUCLEONICS SERIES	
620.5 A3130	AKADEMIIA NAUK SSSR. IZVESTIIA. MEKHANIKA	
620.5 A3130	AKADEMIIA NAUK SSSR. IZVESTIIA. MEKHANIKA I MASHINOSTROENIE	
620.5 A3130	AKADEMIIA NAUK SSSR. IZVESTIIA. MEKHANIKA ZHIDKOSTI I GAZOV	
620.5 A3130	AKADEMIIA NAUK SSSR. IZVESTIIA. OTDELENIE TEKHNICHESKIKH NAUK.	
620.5 B346	BAUINGENIEUR	DR
620.5 EN5	AMERICAN JOURNAL OF MINING, ENGINEERING, GEOLOGY, MINERAOLOGY,	STR

FIGURE 10-8 Serials listing—departmental libraries—architecture

03/18/69 PAGE 2

AC. INTERNATIONAL ASBESTOS-CEMENT REVIEW	N9,1958-
AIA JOURNAL	V1,1944-
ASPO NEWSLETTER	V16,1950-
ABITARE	N52,1967-
AMERICAN ARTIST	V20,1956-
AMERICAN BUILDER	V50,1930-
AMERICAN CITY	V27,1922-
AMERICAN CONCRETE INSTITUTE. PROCEEDINGS	V42,1946-V45,1949
AMERICAN HERITAGE	V20,1968-
AMERICAN INSTITUTE OF ARCHITECTS. BULLETIN	V1,1947-V11,1957
AMERICAN INSTITUTE OF PLANNERS. JOURNAL	V24,1958-
AMERICAN SOCIETY OF ARCHITECTURAL HISTORIANS. JOURNAL	V1,1941-V4,1944
AMERICAN SOCIETY OF CIVIL ENGINEERS. PROCEEDINGS	V48,1922-V76,1950
AMERICAN SOCIETY OF CIVIL ENGINEERS. TRANSACTIONS	V4,1875-V124,1959
AMERICAN SOCIETY OF CIVIL ENGINEERS. STRUCTURAL DIVISION. JOURNAL	V82,1956-
AMERICAN SOCIETY OF PLANNING OFFICIALS. PLANNING. SELECTED PAPERS	1943-
ANTIQUES	V94N6,1968-
ARCHITECT AND BUILDING NEWS	V193,1948-V200,1951
ARCHITECTS' AND BUILDERS' MAGAZINE	V32,1899-V43,1911
ARCHITECTS' JOURNAL	V102,1945-V133,1961
ARCHITECTURAL DESIGN	V17,1947-
ARCHITECTURAL FORUM	V26,1917-
ARCHITECTURAL INDEX	1950-
ARCHITECTURAL RECORD	V1,1891-
ARCHITECTURAL RECORD OF DESIGN AND CONSTRUCTION	V16,1946.
ARCHITECTURAL REVIEW	V71,1932-
ARCHITECTURAL SCIENCE REVIEW	V5,1962-
ARCHITECTURE AND BUILDING	V46,1914-V62,1930
L'ARCHITECTURE D'AUJOURD'HUI	V4,1934-
ARCHITECTURE IN AUSTRALIA	V57,1968-
ARCHITECTURE IN GREECE	V2,1968-
ARCHITEKTONIKI	V13,1959-
ARCHITEKTUR-WETTBEWERBE	N19,1957-
L'ARCHITETTURA	V10,1964-
ARENA	V81,1965-V83,1967
ARENA/INTERBUILD	V83N915-23,1968.
ARKITEKTEN. TIDSCHRIFT FOR BYGGNADSKONST	1950-1960
ARKITEKTUR (COPENHAGEN)	N2,1958-
ARKITEKTUR (STOCKHOLM)	N1,1959-
ARKKITEHTI-ARKITEKTEN	N1,1952-
ARQUITECTURA	N11,1942-
ART BULLETIN	V50,1968-
ART IN AMERICA	V46N3,1958-
ART INDEX	V6,1944-
ART INTERNATIONAL	V8,1963-
ART JOURNAL	V27N3,1968-
ART NEWS	V47,1948-
ARTFORUM	V7N5,1969-

CATALOG REPORT GENERATOR OF THE SYSTEM

The requirement of the program is selection of desired information from the master file by means of the data control cards.

Inputs of the program are (a) the master file of serials records and (b) a data control card for each item of information that is to appear in a given report.

The output of the program is a magnetic-tape file containing information for each of the requested reports, or catalogs, or both.

Functions of the program are (a) stripping from the master serials record file the data required for the printing of catalogs or special reports as specified by the data control cards and (b) formatting such information and transferring the resulting record to an output work tape.

SORT PROGRAM OF THE SYSTEM

The requirement of this program is to sort each of the report work tapes into desired sequence specified by the sort control card.

Inputs to the program are (a) the report work tapes created by the catalog report generator program and (b) the sort control cards that define the order in which records are to appear.

The output of the program is a file of the records to be printed as a report or catalog.

UTILITY UNIT PROGRAM OF THE SYSTEM

The requirement of the program is to print in desired format any standard catalogs or special reports.

Inputs to the program are (a) the sorted report file and (b) the print control card that defines the format in which the information is to appear.

The output of the program is the printed catalog or report.

DESIGN OF COMPUTER SYSTEMS—OPERATING SYSTEMS

So far we have described in some detail the systems for compiling the original input to the serials system for setting up the master record file and for correcting this file and for setting up the reporting system required to produce the necessary reports or catalogs. The primary purpose of the serials operating systems is to provide efficient and accurate control of all records and the transactions.

Figures 10-9 and 10-10, systems charts of the monthly and weekly operating systems, outline the programs and their interrelationships that are necessary to fulfill the following requirements:

1. Guarantee that all issues not received by the library will be claimed promptly and on a set schedule and that the renewal of all serials will be controlled by the system.

2. Predict the time of receipt of the journal thereby establishing a positive control over the timely receipt of all issues.

3. Maintain and control the accounting records necessary for the serials system internally as well as report to the comptroller the information required for payment of invoices.

4. Provide the binding department with the necessary information and records for those titles that are to be bound.

5. Handle the many changes that occur in the various serials titles.

Outputs of the operating systems are (a) an updated master file of all serials records; (b) punched control cards anticipating receipt of issues; (c) reports for controlling claiming, renewals, and binding; and (d) statistical and accounting reports.

The operating systems work within three time frames: monthly, weekly, and daily. We shall now review functions that are performed by both the computer and the clerical.

MONTHLY OPERATING FUNCTIONS

The monthly run of the check-in program (Figure 10-9) is the key to the successful fulfillment of the operating system. It is the function of this program to forecast the transactions that are to occur within the serials system for the coming month and to produce the reports and control cards that are needed to handle successfully these transactions. The monthly run of the check-in program has specific functions that must be performed.

1. A forecasting must be made of the issues to be received for each of the active titles during the coming month. Utilizing the FREQ (frequency) data in conjunction with the months of publication data (JFM . . . field) the program will determine whether an issue is to be received or claimed; the check-in program will determine, when possible, the volume

FIGURE 10-9 Systems chart—operating systems—monthly run of the Check-in program

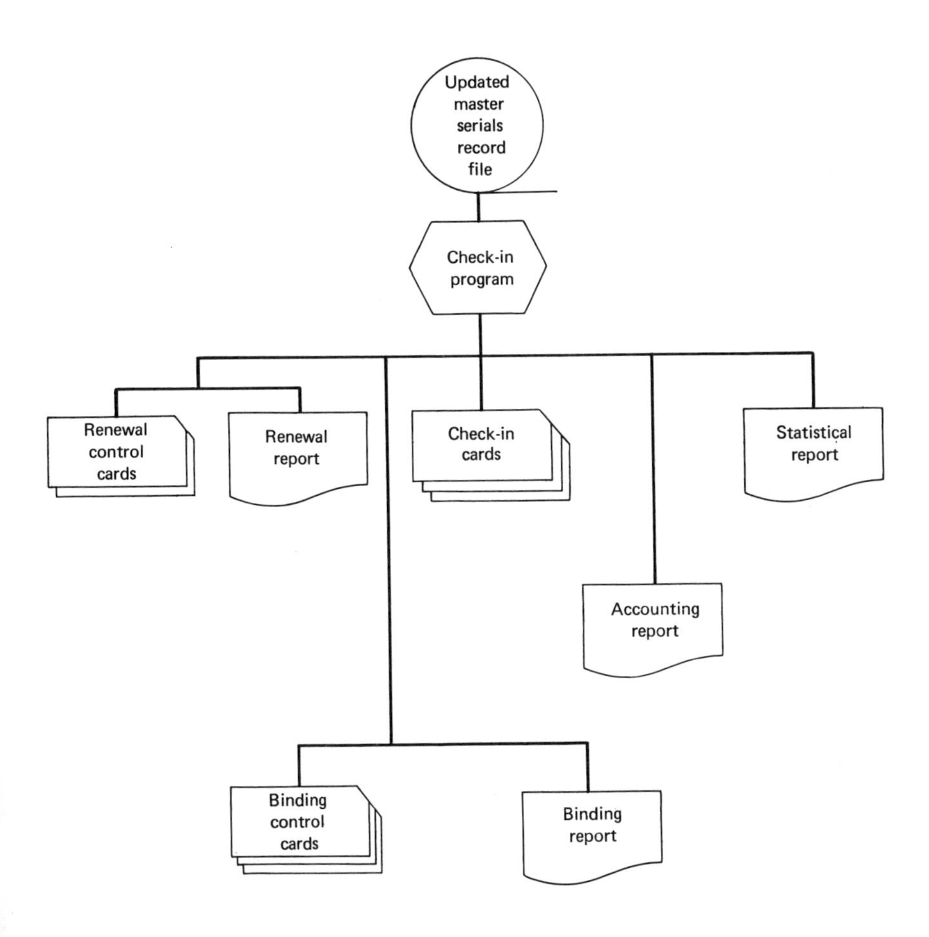

and issue numbers to be received in the coming period. A control card will be punched for each issue to be received. Simultaneously the program will post in the waiting receipt and claims section of the master record the control data for each particular issue. The check-in cards will be filed in the serials department awaiting receipt of the issue.

2. A listing must be produced of all titles whose subscriptions are to be renewed during the second month following the monthly run. (For example, during the June run the computer will forecast those titles to be renewed during August of that year.) This program determines the renewal date by utilizing the date of last renewal found in the DATE SUBS START (date subscription start) field and the period of the last subscription from the SUBS LGT (subscription length) field, both found in the accounting data section of the master record. The renewal report is produced by vendor or by publisher showing the titles that should be renewed in the period. The report contains the call number, the title, the last renewal date, the subscription length, and the estimated price of renewal. At this point, the program also produces a control card for each item being renewed. The control card contains the call number and short title. The program also posts to the accounting data control section the renewal cost incurred.

3. A binding report is necessary for those titles with sufficient consecutive issues to warrant binding into a physical volume. A control card is produced for each volume to be bound and an indication is made in the binding control section of the master record that an order has been issued for binding.

4. Accounting reports are required for the financial control of the system. These reports are based on encumbrances incurred by each fund and department account. In order to obtain a net balance for each fund and department account at the end of each month the encumbrances are adjusted to actual cost on receipt of invoices.

5. The program must generate monthly statistics of the number of new titles added and their cost, the number of renewals and the estimated cost of renewals, the number of issues received, the number of bound volumes added to the serials collection, and other statistical information.

For all titles whose frequency of receipt is irregular the check-in program will, on the receipt of an issue, produce a check-in card for the next arrival. This card contains only the call number and the short title of the issue and, if possible, the volume and the issue number but not the month or year.

DAILY OPERATING FUNCTIONS

This section describes the functions of the staff of the serials system and the data processing operator in maintaining records.

1. *Check-in*: When an issue is received, the serials clerk removes from the punched-card file the corresponding check-in card for the particular volume and issue. In the case of an irregular publication it is necessary to record on the check-in card (if not there already) the volume and issue number, adding the month and year of publication. The clerk marks the call number on the issue and checks for changes such as title and frequency in the journal. If variations are not found, the check-in card is forwarded to the data processing operator. If changes have occurred, they are recorded on the serials control record worksheet, indicating the call number and the changes in the appropriate fields. The check-in card and the worksheet are forwarded to the data processing operator who, using the auxiliary punched-card reader attached to the paper-tape typewriter, enters the check-in card data into the paper tape. The call number, issue, month, and year are punched into the paper tape as input to the weekly run of the check-in program. If corrections or changes to a given title are made, the data processing operator holds these until the end of the day or week when the correction tape for input to the translate and edit program is produced.

2. *Ordering of New Serial Titles*: On receipt of the first issue of a new title the serials librarian prepares a serials control record worksheet that is forwarded to the data processing operator for entry of the new title into the system.

3. *Renewal of Subscriptions*: The renewal control cards and the renewal reports are received from the computer. The renewal punched card is filed. The renewal report is verified and two copies forwarded to the supplier or the publisher for the renewal of the indicated titles. When the invoice is received, the renewal control card for each title on the invoice, the invoice number, and the amount of the renewal are posted to the renewal control card. The renewal control report is checked for each title and the renewal control card is then forwarded to the data processing operator. The operator duplicates the prepunched information on the renewal control card and key punches the new information (invoice date and number and renewal cost) and simultaneously produces a punched paper tape with this information. This is accomplished by using a key punch that is cabled to the paper-tape typewriter.

4. *Claiming of Serial Issues*: The weekly run of the check-in program prints out the claim notices. These notices are reviewed prior to mailing because notice may have been received from the supplier or

publisher that the item will be delayed. In such event the computer record is updated. If the claim is routine, the notice is forwarded to the supplier. On receipt of the claimed issue the check-in card for that issue is removed from the file and forwarded to the data processing operator.

5. *Binding Control*: On receipt of the binding control list and the control cards the binding department uses the control list as its worksheet for the coming month. The journals are taken from the shelves and checked for completeness. Binding orders are then prepared for the completed volumes. It is the responsibility of the binding clerk to submit correction reports for any issues or indexes found missing.

WEEKLY RUN OF THE CHECK-IN PROGRAM

At the end of each week all input paper tapes for the serials system are forwarded to the computer center for entry into the weekly run of the check-in program (Figure 10-10). These tapes are processed through the translate and edit program. The output of the translate and edit program is a magnetic-tape work file that serves as input to the sort program. The output of this sort program, arranged by call number and type of transaction, is the input to the weekly run of the check-in program.

The function of the check-in program is to post all transactions that have occurred to the serials master record file. The primary output of the weekly check-in program is the claim notices. During the weekly run the program audits each record, comparing the date of a forecasted arrival in the WAITING RECEIPT sector of the record to the CLAIM CTL (claim control) date. If the issue is overdue, a claim is issued. If a subsequent published issue is received prior to the receipt of an earlier published issue, a claim is issued for the earlier issue regardless of the claiming schedule. When an issue is received, the WAITING RECEIPT sector of the record is cleared and the issue is posted to the binding control section of the record. The output of the weekly run is the updated serials master control file that is always the source of all printed reports.

CRITICAL COMMENTS AND OBSERVATIONS

A critical operation within the serials system is the check-in and claiming of journal issues. There is an alternative to the punched card, paper-tape check-in and claiming operations in the computer-based serials system. The direct use of the punched card from the check-in file as input to the weekly

FIGURE 10-10 Systems chart—operating systems—weekly run of the check-in program

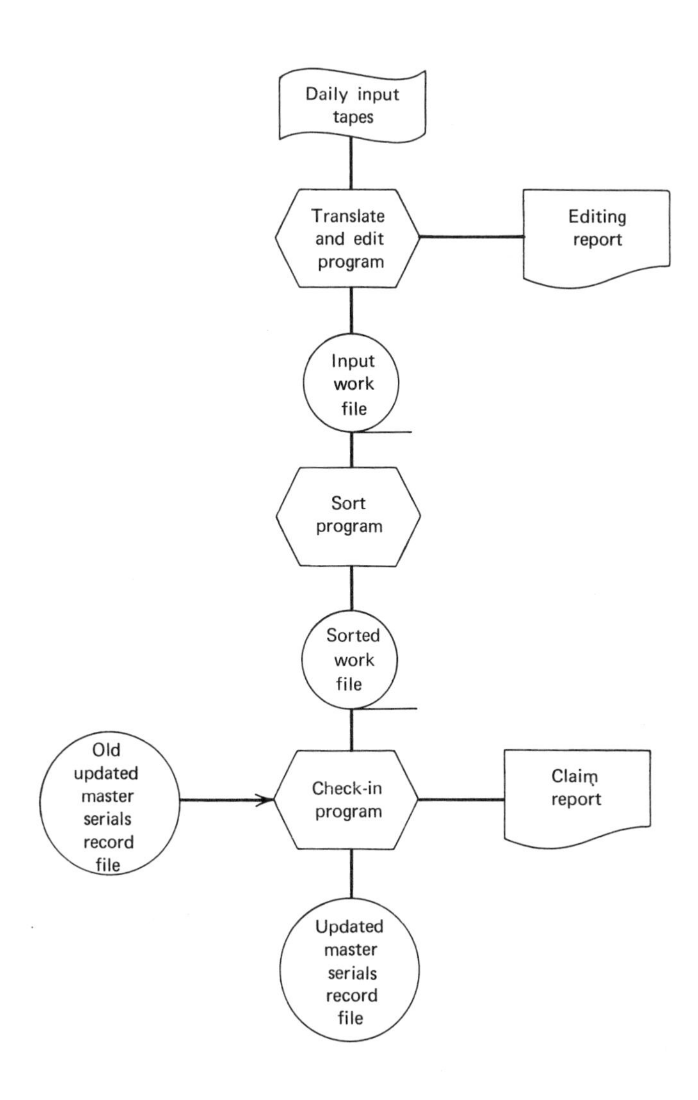

run of the check-in program (Figure 10-10) could be a more efficient way to update the master serials record file. If the punched cards produced by the computer for the check-in operation contained the required information for check-in, the intermediate step of paper tape might well be eliminated. If this were done, no translate program would be necessary because punched cards do not require this. This phase of the weekly program would only require a card-to-tape utility program.

Another alternative could be that instead of punched cards, a monthly printout listing of expected journal issues could be used. After a journal issue is received, it could be crossed off the printed list, marked, and sent to the shelves. At the end of the month the printout would be sent to the claiming subsystem for processing. Those titles not crossed off would require a claims notice. The initial notice, both to the vendor and the computer record, would be produced on the paper-tape typewriter. Subsequent claim notices would be issued by the claims program under computer control. Updating the computer record with information about the issues that have been received would be accomplished by the computer *not* being notified. If no notice were received, programming would provide for the computer record to be updated automatically on the assumption that the issue had been received.

The advantage, if any, of the printout listing over the 80-column punched-card record is mainly in the printout's not having to be concerned with abbreviated titles and other truncated bibliographic data.

chapter eleven

TOTAL SYSTEMS CONCEPTS IN THE DESIGN OF A COMPUTER-BASED CIRCULATION SYSTEM

As implied in the heading of this chapter total systems concepts are viewed and evaluated to indicate their importance in designing a computer-based circulation system. Planning with total systems concepts in mind should disclose most of the ramifications that will occur after implementation of a new system.

TOTAL SYSTEMS DEFINED

The phrase "total system," as used here, concerns itself with more than the immediate library system or the overall parent administrative system. Rather it is meant to designate the symbiotic relationships within the library and the general relationships of the library with society; that is, it is meant to deal with the interrelationships or interdependence among systems that affect one another and in many instances justify and facilitiate the existence and performance of systems. Interdependence is the key word describing the inner workings of systems within a total system. A limited, two-dimensional representation showing one phase of interdependence of systems is illustrated in Figure 11-1, Total System. This diagram illustrates some of the interrelationships that do exist. General society, all inclusive, is the source not only for the people who are to manage the library but also for the library users and the supply of outside supporting services and materials to the library. When viewed in closer perspective, the interdependent systems of the library in Figure 11-1 become obvious. Stylized relationships between the systems within the total library are shown encompassed by the administration and planning system. Circulation is depicted as logically interrelated with the reference system and the cataloging system. From the latter newly processed library materials are received and controlled by the circulation system in order to provide the library's users with ready access. The location symbols assigned and indexes prepared and maintained by cataloging are used by the circulation system and by the library users to locate and relocate desired items. One of the relationships between acquisitions and cataloging in which the one furnishes the material for the other to process so as to make it readily accessible to the library user and to other systems, further serves as an illustration of the total systems concept. Reciprocally the records that cataloging maintains allow acquisitions to search each purchase request effectively before ordering to avoid unwanted duplication of material and work.

DESIGN OF CIRCULATION SYSTEM

When the circulation system is accepted as a system that is directly interdependent with other systems within the library as well as with outside systems (that is, the user), a more lucid understanding of the place of the circulation system within the total library system may be realized. It should also become increasingly clear that the design of a new system with its stated demands and requirements and its inputs/outputs will have direct bearing in varying degree on all the other systems within the total system. Therefore it becomes very important in designing a new system to be fully aware of what immediate and long-term effects a new printed form, method or procedure will have on existing cosystems.

In considering the design of a computer-based loan system it should at least be able to emulate the worthwhile elements of the existing system. These elements, probably common to all circulation systems, include the ability to:

1. Indicate what book is out and who has it.

2. Indicate the date the book was taken out and the date it is to be returned.

FIGURE 11-1 Total system

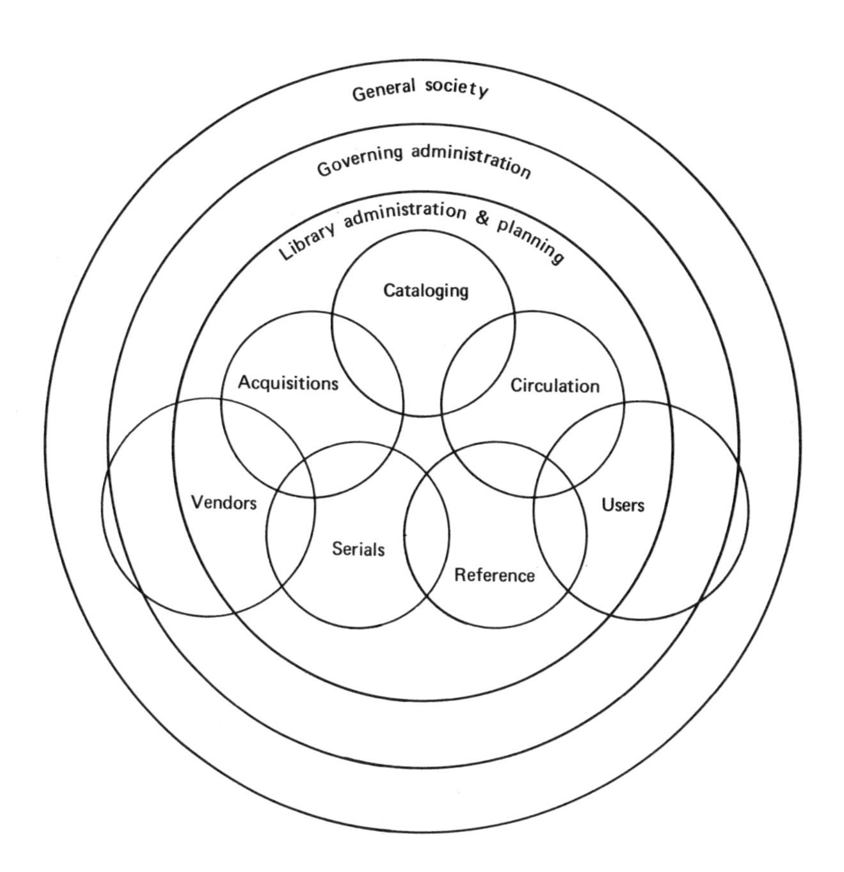

3. Allow for a daily count of circulation activity by type of material and possibly by the status of the user.

4. Search for missing books, and if found or not, notify the requestor of the book's availability.

5. Allow for reserving charged-out library material.

6. Charge out and discharge library material.

7. Maintain records of overdue books and issue notices as required.

8. Allow for a daily updating of the master file of outstanding circulation by interfiling the daily transactions.

9. Provide on users' notices the call number, title, and due date of the requested material.

10. Enable the library to know the user's identify prior to borrowing any material.

11. Facilitate recall of charged-out material prior to its due date.

Accepting the necessity for maintaining the objectives of the existing circulation system, the computer-based system could be designed by the analyst to gain a variety of improvements in the maintenance and accessibility of circulation records and more meaningful statistical records.[1] In addition to improving on present objectives the computer-based circulation system should be designed to be flexible enough to:

1. Answer a user's enquiry as to the status of a book (out, in, missing) by current printouts arranged variously by author or title, call number, and borrower, including in each listing the date the material was charged out and is to be returned.

2. Ascertain the user's identity and be able to correlate by statistical analysis, his subject background with the type of material he charges out.

3. Provide work-unit statistics to indicate hourly and daily workloads.

4. Maintain a historical file of loan transaction records for use in periodic statistical studies.

5. Be as mechanical as possible in its charge and discharge operations so as not to be dependent to any appreciable extent on manual keying or writing functions and thus make charge and discharge operations as rapid and free of error as possible.

6. Allow for *portable* terminal data collection devices for use in inventorying the book collection and perhaps to allow for user self-charging of material.

[1] For a discussion of the circulation system's role in the total library system see: R. T. Kimber, *Automation in Libraries* (International Series of Monographs in Library and Information Science, Vol. 10) Pergamon Press, London, 1968, Chap. 4, Circulation Control, pp. 49-74.

7. Take into account short circulation periods, anywhere from a few minutes to several hours, allowing for class reserve books to be controlled by the computer-based circulation system (This may be best performed in an on-line system).

8. Provide for continuing analysis of the book collection by frequency of use by subject and subject background of user so as to evaluate collection growth and adequacy.

9. Indicate the retrospective and current number of uses of any book within the library's collection to facilitate the ordering of additional copies of materials or the weeding out of inactive material.

10. Permit expeditious inventorying of the collection with machine-readable records.

11. Maintain statistical records of user requests for missing books.

12. Allow for conversion to an online circulation system.

13. Be able to generate on request lists of books charged to individual users.

14. Analyze the use made of class reserve reading material.

15. Machine produce on request and as needed recall and overdue notices.

Not only should the systems analyst be aware of what objectives the circulation system should fulfill, he should also be able to evaluate the commercially available mechanical data collection (input) devices best suited to realize the total systems concepts in the computer-based circulation system. Along with the decision made regarding the type of input device to be used to capture the transaction record, a master punched book card (the input) for each circulating item, probably an 80-column punched card, should be produced with enough information to identify the book. If the bibliographic data is already computer based, it is only a matter of stripping the necessary information from the existing computer record and producing the punched book card, interpreting the Hollerith code to make the card man-readable, and inserting it into the book. An example of planning for the future is provided in the case of the library system that has used no book cards, even under the manual system, but still provided a book card pocket in the back of each book in the planned event that punched cards would be used as master book cards in a computer-based circulation system.

If the cataloging system is at present converting its records to machine-readable form, the master book card could be an economical by-product of the conversion process. As the shelf list is converted selected information would be punched into the punched card to create a master book card for the circulation system. If no machine-readable bibliographic records exist, the conversion of records to machine-readable form to produce the master book cards is a costly and critical operation prior to the implementation of the computer-based circulation system.

MASTER BOOK CARD

Provided that a machine-readable record can be produced, the analyst should define the data to be carried on each master book card. He has less than 80 columns with which to uniquely identify the book because some of the card space is allocated for the identification number of the user, which will appear on the transaction card combined with the data from the master book card. Therefore the information contained in the master book card and that added to the transaction card, although minimal, should allow for:

1. Positive identification of the specific book by the master book card. The order or accession number or the full call number, or both, including volume, part, copy, and so forth, would appear to be necessary.
2. Enough bibliographic information to enable the user and the library staff to identify the book by author or title, or both, in a printout list of loan transactions. It may not be necessary for this information to be punched into the master book card because a machine-readable shelf list in computer storage can be used either in real-time (immediate access) or in an off-line situation for complete identification of the book if the call number or the order number is used as the key to the entire bibliographic record.
3. The due date; the computer program can take this into account and provide the due date on the printout.

An advantage to creating a master book card that is bibliographically self-sufficient is that it allows for generation of computer printouts directly as requested. The printing out of the outstanding circulation file can be accomplished by sorting and revising only the information pertinent to circulation without referencing the entire store of cataloging data.

USER IDENTIFICATION CARD

The data to be carried by the machine-readable user identification card should be specified by the analyst. If properly designed, this card serves as an excellent example of application of the total systems design concept. For the purposes of the library the identification card would be Hollerith coded with some identifying number that would serve as the key to each user's name and address and other pertinent information in a separate machine-readable file. This key number, which would be punched into the user's card, should have certain characteristics:

1. It should be short enough for the identification card and for the remaining space on the master book card when combined in a transaction card.

2. It also should take into account the accessibility of already existing machine-readable borrower files or other machine-readable files carrying this type of information that could be used to provide addresses and other data for collection development studies. If no machine-readable files exist, it is a matter of converting manual borrower or user files.

An identification card to be carried by each user should be provided or an existing card modified to meet the demands of the new system. This card requires key punching and if this is not amenable to an already existing card, a new one will be needed. If a new identification card is required, it should be designed to carry (a) positive identification of the user by a photograph, name, the Hollerith coded identification number, and possibly the person's signature and (b) any additional information that is compatible with other uses of the card outside of the library.

In support of the total systems concept the identification card should be designed to be useful, if possible, in other systems. By embossing the name and birthdate of the carrier on the card it can be used as a "charge plate"; and by providing a photograph and signature it can serve as a very compact general-use identification card. In a university situation such a card would find application in library, bookstore, registration, and student activities transactions. The most important characteristic of identification number to be punched into the card is the adaptability it will or should have in existing machine-readable identification files in various systems. Therefore special consideration should be given to the machine-readable number to be used: is it to be a locally unique number or a nationally unique one like the Social Security number?

In support of the nationally unique number the implications of the increasing interest and activity in shared library resources indicate the need for universal identification. For example, in a network of libraries if a number of the cooperating libraries implement computer-based circulation systems and users from any one of these borrow books from various libraries in the system, the identification number should be machine-compatible throughout the system. This obviously necessitates a machine-readable central address file accessible to the cooperating libraries.

CONTRIBUTION TO LIBRARY GOALS

The format and content of the master book card and user identification card are important to the contribution of the circulation system to the total library systems goals. This is in terms of the circulation system's ability to contribute effectively to the administration of the library and to collection development for best service to users. If a library is to capitalize on the statistical information available from daily circulation transaction records, some planned means of making use of this data should be provided. The materials circulated are obviously an indication of user subject and intellectual interests and development. A variety of statistical records can be supplied by a functionally designed computer-based circulation system. Continuing statistical analysis can serve to signal and predict fluctuations in demands being placed on the library. Therefore data collection as a by-product of the circulation control system should be designed to be sensitive to the subtle changes in user requirements and expectations. The result of such an approach will verify the aptness of the techniques already in use to attain the library's service goals or it may show the conventional techniques to be inadequate.

In order to analyze properly the information gathered the data collection function needs to be designed to (a) incorporate flexibility and compatibility in the gathered data to generate needed reports; (b) balance economic limitations against statistical reports clearly required in terms of the broadness of their application; and (c) consider the needs and demands of the user, the types of users, and the ecology of the library. A library network's requirements may well differ from those of any one member library.

DATA COLLECTION DEVICE

The mechanical input device in a computer-based loan system should be able to generate the necessary data for computer input. Apart from expeditiously accepting and processing the inserted master book card and the user's card, the machine should be able to withstand extensive use in repeated charging and discharging functions. It would seem that the more simple the machine's construction and the more independent it is of auxiliary equipment, the less chance there is of it failing. It should have the feature of being able to go on line. Regarding the routine operations in which the machine will be involved, the discharging of books may be a problem area under certain conditions. In a one-card system the master card is used to produce the charge and discharge record for the input to the computer and in doing so requires that the book always stay with the master card. As a solution to this problem a two-card system produced by the input device is suggested. Two cards are produced from the master book card: (a) a charge-out card that contains the same data as the transaction card and is placed in the book pocket along with the master card and (b) a transaction record card that serves as computer input. By this method the charge card can be removed from the returned book and the book sent immediately to the shelves. The charge-out card serves to clear the computer record by matching with the transaction record as it appears in the computer.

STATISTICAL REPORTS

By defining the necessary data to be incorporated in each master book card and user card the analyst should be able to extract a number of valuable statistical use studies and measures from the loan record. For example, because there are no precise means of ascertaining stack or in house use of books, those charged out should have significance in measuring the use of the collection and in collection development. Because the computer can readily provide records of actual use and nonuse of books, this has direct bearing on collection development by the reference system. If a subject represented by a group of books is no longer referred to, the purchase of new materials in that subject field may be questioned. Effective collection development through awareness of use should also provide data for subject profiles of specialized users, augmented and confirmed by personal interviews. If circulation records reveal that a title is used extensively, additional copies may be automatically purchased.

Within the cumulative circulation statistical record, book use can be measured and classified by call number and perhaps, preferably, by means of the subject "tracings" in the computerized shelf list correlated with the subject background of users. From observation of circulation trends suggested reading lists of currently appearing titles could be produced as an awareness service in the subject areas of readers' interests. Through the use of statistics it would be possible to predict inductively by any subject's use or nonuse the significant growth or decline of interest by library users. This could be a direct aid in the library's book-selection subsystem.

In most, if not all libraries, the administration is in competition for increased budgets with other units and offices of the governing administration. The library administration should be able to demonstrate through quantitative and qualitative use records of the library's collection the need for an increased book budget. These use studies may also indicate need for a decrease in or shifting of acquisitions funds that may have been wanted in maintaining unused subject collections.

An additional operation that would be dependent on a computer-based circulation record and shelf list would be that of inventorying the collection entirely or on a sampling basis. Records of user queries for missing books in a given area might indicate the need for an inventory of that area.

EVALUATION AND WEEDING OF LIBRARY COLLECTION

The weeding of library collections is often neglected. By comparing a computer listing of materials used in the last five years, for example, against a machine-readable shelf list, a list of material *not* loaned in that period could be produced. This list could serve as the basis of a weeding operation. This appears to be a better method than basing reasons to discard materials or place them in inactive storage on general age-use mathematical models. If as one of the primary steps in implementing a computer-based system all master book cards are prepared but not inserted into the books until they are charged out, the remaining file of master book cards not inserted after *X* number of years would indicate those books not circulated in the predesignated period of time. This remaining file could be processed to produce a list of books eligible for weeding operations—either withdrawal or referral to inactive storage.

As implied previously circulation transaction records are one of the best sources of information about collection-use patterns. Other means of ascertaining use, including on-site observation and the counting of unshelved books on tables, may be too limited and may not cover the overall aspects that an extensive collection of computerized circulation records could provide. The percentage of books ten years old or older found on study tables in a two-week period and used as the basis to extrapolate a general statement on the entire collection's use does not necessarily provide sufficient guidance for weeding of a collection. Under the total-system concept and the management by exception principle all facets of collection use are considered or may be considered. This reveals not only the extent of use of older books but also an itemized listing of these books, their subjects and where applicable, the subject background of their users.

By exploiting the full capability of loan statistics to measure and project library use, service to various groups of library users may be improved. The current loan transaction indicates what is happening and thereby may allow the library's administration to respond to the demands of the library's users as deduced from the loan statistics. If a definitive pattern can be obtained about how effective the library is in meeting user requirements, far more precise parameters of library service may be established.

A further illustration of the total systems concept is the loan transaction record accumulated over a period of time; this could be very meaningful in arriving at staff subject specialist requirements. The nature of the user, derived from subject profile analyses by questionnaire and personal interviews as well as from actual use reports, would provide a guide to the subject background and experience needed for meeting the needs of the library's community. Although this would be only a partial measurement, it would still contribute to a broader understanding of job requirements.

ONLINE APPLICATIONS

The analyst in designing a system utilizing the total systems concept not only should be aware of what effect the design will have on other systems but also should be cognizant of and prepared for new technological developments. A computer-based circulation system should be designed to take into account on-line operating procedures. By an on-line operation is meant direct input to the computer, usually with immediate response from the computer. Printouts may still be used as guides to the outstanding circulation. An on-line procedure makes unnecessary the preparation of the computer input, the transaction card of the master book card's data combined with the user's identification. This transaction data goes directly into the computer storage. In an on-line application with immediate access, loan information will not only go directly to the computer but also, as requested, will be printed out on a typewriter terminal or displayed on a cathode-ray tube.

BIBLIOGRAPHY

This is a selective bibliography of topics about or related to library systems analysis. It is an alphabetically arranged list of journal articles, selections from symposia, books, and parts of books. A bibliography of pertinent bibliographies is found at the end.

All of the items included have been reviewed by the authors excepting a few of those that were unobtainable.

While the emphasis of this bibliography is library oriented and library literature is frequently cited, business management literature sources have been referred to as well since business techniques may readily be adopted for library management. Further, management literature is cited because the library profession seems to have had very little to say on the systems approach in library management.

The time covered by this bibliography is from 1954 to May, 1969. *Library Literature* was surveyed from the 1955 cumulation to the present. References and bibliographies found in works consulted and reviewed were culled for their pertinency for inclusion.

The numerous case studies of automation and information retrieval in libraries in general have not been included. The few that are cited are considered to contain meaningful observations about the state of the art and not exclusively factual descriptions of how something is or was done.

Journal abbreviations conform with usage in *Library Literature.*

It is the belief of the authors that selective reading of items in this bibliography should lead to a good understanding of the applicability of systems analysis to library operations. The January 1954 issue of *Library Trends* on scientific management in libraries is suggested as a sound starting point in a reading program.

The subjects of this bibliography may be categorized under six general headings: library systems analysis, user requirements, flow charting, technical writing, automation in libraries, and electronic data processing.

Subjects beneath library systems analysis (Burkhalter, Becker, Hayes, Taylor) include scientific management (Dougherty), forms design (Neumaier), business management with systems concepts (Optner, Neuschel, Systems and Procedures Association), techniques of measuring library effectiveness (Schultheiss, Pings, Lamkin, Dougherty, Drott, Kozumplik), costing (Herner), job descriptions of the systems analyst (Minder, Covill), and conflict and opposition to change in libraries (Bundy, Williams).

Although the study of *user requirements* is especially important in the total library system, there is not very much literature about the methods and techniques of obtaining user-requirement data. The making of library surveys (Line) and studies about user-library interchange (Taylor, Trueswell, Cammack) are topics that have been included in this bibliography.

Flow charting, both computer (Flowcharting Techniques) and manual (Mullee, Gull), is set apart from the other techniques of analyzing systems because of its unique effectiveness in systems evaluation and design and in in-service training of staff.

Skill in technical writing (Wyle, Ulman, Kapp), which is important in preparing required reports in systems study, is another significant topic.

Automation in libraries includes a variety of subheadings including, costing (Fasana), evaluatory criticism of progress made in the field (Bryan, Stuart-Stubbs, Strain), the future (Wasserman, Batty), computer-based library systems (DeGennaro, Kilgour, Parker, Davie, Kimber), and staffing requirements (Gull, Herner).

Under the heading of electronic data processing appear such topics as introductory texts to automatic data processing (Arnold, Wainwright, Brandon, Gregory, Greenwood), total systems (Meacham), punched cards systems (Levy), information handling (Bourne, Becker), computer programming (Davis), and computers in education (Caffney).

1. Adelson, Marvin, The System Approach: A Perspective, *Wilson Lib. Bul.*, **42**:711-715 (1968).
2. Adkinson, Burton W., Trends in Library Applications of Data Processing, in *Clinic on Library Applcatons of Data Processng, Proceedings, 1963,* ed. Herbert Goldhor, University of Illinois, Graduate School, Distributed by the Illini Union Bookstore, Champaign, Ill., 1964, pp. 1-8.
3. American Association of School Administrators, *EDP and the School Administrator,* The Association, Washington, D.C., 1967.
4. American Management Association, *Establishing an Integrated Data Process-*

ing System: Blueprint for a Company Program, The Association, New York, 1956.

5. American Management Association, Administrative Services Division, *Advances in EDP and Information Systems* (AMA Management Report No. 62), The Association, New York, 1961.
6. American Management Association, Finance Division, *Data Processing Today* (AMA Management Report No. 46), The Association, New York, 1960.
7. American Management Association, Office Management Division, *Organizing for Effective Systems Planning and Control,* The Association, New York, 1956.
8. Andrews, Theodora, The Role of Departmental Libraries in Operations Research Studies in a University Library, Part 1, Selection for Storage Problems, *Sp. Lib.,* **59**:519-524 (1968).
9. Andrews, Theodora, The Role of Departmental Libraries in Operations Research Studies in a University Library, Part 2, a Statistical Study of Book Use, *Sp. Lib.,* **59**:638-644 (1968).
10. Arnold, Robert R., *et al, Introduction to Data Processing,* John Wiley, New York, 1966.
11. *Automation in the Library—When, Where and How,* ed. Theodora Andrews with the assistance of Molete Morelock, Purdue University, Lafayette, Ind., 1965.
12. Awad, Elias M., *Automatic Data Processing: Principles and Procedures,* Prentice-Hall, Englewood Cliffs, N. J., 1966.
13. Bare, C. E., Conducting User Requirement Studies in Special Libraries, *Sp. Lib.,* **57**:103-106 (1966).
14. Batty, C. D., ed., *Libraries and Machines Today; a Sequel to the Library and the Machine,* North Midland Branch of the Library Association, Scunthorpe (Lincs.), 1967.
15. Batty, C. D., ed., *The Library and the Machine,* North Midland Branch of the Library Association, Scunthorpe (Lincs.), 1966.
16. Batty, C. D., The Library of Tomorrow, in *The Library and the Machine,* ed. C. D. Batty, North Midland Branch of the Library Association, Scunthorpe (Lincs.), 1966, pp. 37-52.
17. Becker, Joseph, Communications Networks for Libraries, *Wilson Lib. Bul.,* **41**:383-387 (1966).
18. Becker, Joseph and Robert M. Hayes, *Information Storage and Retrieval: Tools, Elements, Theories,* John Wiley, New York, 1963.
19. Becker, J., System Analysis, Prelude to Library Data Processing, *ALA Bul.,* **59**:293-296 (1965).
20. Bellomy, F. L., The Systems Approach Solves Library Problems, *ALA Bul.,* **62**:1121-1125 (1968).
21. Boaz, Martha, Evaluation of Special Library Service for Upper Management, *Sp. Lib.,* **59**:789-791 (1968).
22. Bocchino, William A., *A Simplified Guide to Automatic Data Processing,* Prentice-Hall, Englewood Cliffs, N.J., 1966.

23. Bolles, Shirley W., The Use of Flow Charts in the Analysis of Library Operations, *Sp. Lib.*, **58**:95-98 (1967).
24. Bourne, Charles P., *Methods of Information Handling,* John Wiley, New York, 1963.
25. Bourne, Charles P., Some User Requirements Stated Quantitatively in Terms of the 90 Percent Library, in *Electronic Information Handling,* ed. Allen Kent and Orrin Taulbee, Spartan Books, Washington, D. C., 1965, pp. 93-110.
26. Brandon, Dick H., *Management Standards for Data Processing,* Van Nostrand, Princeton, N.J., 1963.
27. Brownlow, J. L., Cost Analysis for Libraries, *DC Lib.*, **31**:54-60 (1960).
28. Brutcher, Constance, *et al,* Cost Accounting for the Library, *Lib. Resources and Tech. Serv.*, **8**:413-431 (1964).
29. Bryan, Harrison, American Automation in Action, *Lib. J.*, **92**:189-196 (1967).
30. Bundy, M. L., Conflict in Libraries, *Coll. & Res. Lib.*, **27**:253-262 (1966).
31. Bundy, M. L., Decision Making in Libraries, *Ill. Lib.*, **43**:780-793 (1961).
32. Burck, Gilbert, *The Computer Age and Its Potential for Management,* Harper and Row, New York, 1965.
33. Burkhalter, Barton R., ed., *Case Studies in Systems Analysis in a University Library,* Scarecrow Press, Metuchen, N.J., 1968.
34. Caffrey, John and Charles J. Mosmann, *Computers on Campus,* American Council on Education, Washington, D.C., 1967.
35. Cahill, B. F., How to Use a Work-count, *Supervisory Management,* **8**:25-28 (1963).
36. Cammack, Floyd and Donald Mann, Institutional Implications of an Automated Circulation Study, *Coll. & Res. Lib.*, **28**:129-132 (1967).
37. Carroll, P., We Need Work Measures, *Sp. Lib.*, **50**:384-387 (1959).
38. Chamis, Alice Yanosko, The Design of Information Systems; The Use of Systems Analysis, *Sp. Lib.*, **60**:21-31 (1969).
39. Chapin, E., Administrative and Economic Considerations for Library Automation, in *Clinic on Library Applications of Data Processing, Proceedings,* 1967, ed. Dewey E. Carroll, University of Illinois, Graduate School of Library Science, Urbana, Ill., 1967, pp. 55-69.
40. Clapp, Verner W., Automation and Data Processing, in *First Governor's Library Conference, Proceedings,* State of New York, Albany, N.Y., 1965, pp. 44-46.
41. Clapp, Verner W., Closing the Circuit; Automation and Data Processing for Libraries, *Lib. J.*, **91**:1165-1171 (1966).
42. *Clinic on Library Applications of Data Processing, Proceedings, 1963,* ed. Herbert Goldhor, University of Illinois, Graduate School of Library Science, Distributed by the Illini Union Bookstore, Champaign, Ill., 1964.
43. *Clinic on Library Applications of Data Processing, Proceedings, 1964,* ed. Herbert Goldhor, University of Illinois, Graduate School of Library Science, Distributed by the Illini Union Bookstore, Champaign, Ill., 1965.
44. *Clinic on Library Applications of Data Processing, Proceedings, 1967,* ed.

Dewey E. Carroll, University of Illinois, Graduate School of Library Science, Urbana, Ill., 1967.

45. Connor, J. M., Management Methods in Libraries; a Symposium—Office Machines and Appliances, *Med. Lib. Assn. Bul.*, **49**:534-540 (1961).
46. Cornell University Libraries, *System Requirements*, The Libraries, Ithaca, N.Y., 1965.
47. The Costs of Data Processing in University Libraries, *Coll. & Res. Lib.*, **24**:487-495 (1963).
48. Covill, George W., Librarian + Systems Analyst = Teamwork?, *Sp. Lib.*, **58**:99-101 (1967).
49. Cox, N. S. M., *et al, The Computer and the Library*, Archon Books, Hamden, Conn., 1967.
50. Culbertson, D. S., New Library Science: a Man-machine Partnership, *PNLA Q*, **29**:25-31 (1964).
51. *Data Processing in Public and University Libraries*, ed. John Harvey, Spartan Books, Washington, D.C., 1966.
52. Davie, C. K., Administration and the Computer: the Context for Libraries, in *The Library and the Machine*, ed. C. D. Batty, North Midland Branch of the Library Association, Scunthorpe (Lincs.), 1966, pp. 4-20.
53. Davis, Sidney, *Your Future in Computer Programming*, Richards Rosen Press, New York, 1969.
54. DeGennaro, Richard, The Development and Administration of Automated Systems in Academic Libraries, *J. Lib. Automation*, **1**:75-91 (1968).
55. Denver University Libraries, *Manual for Time and Motion Studies*, The Libraries, Denver, Col., 1960.
56. *The Development of Objective Measures of Library Functions for Management Decision-making*, State University of New York, Upstate Medical Center, Syracuse, N. Y., 1968.
57. Dougherty, Richard M., Manpower Utilization in Technical Services, *Lib Resources & Tech. Serv.*, **12**:77-82 (1968).
58. Dougherty, Richard M. and Fred J. Heinritz, *Scientific Management of Library Operations*, Scarecrow Press, New York, 1966.
59. Douglass, Paul, *Communication through Reports*, Prentice-Hall, Englewood Cliffs, N. J., 1957.
60. Dow, K. K. W., Tool for Management Evaluation of Library Services, in *Readings in Special Librarianship*, ed. H. S. Sharp, Scarecrow Press, Metuchen, N. J., 1963, pp. 177-188.
61. Drott, M. Carl, Random Sampling: a Tool for Library Research, *Coll. & Res. Lib.*, **30**:119-125 (1969).
62. Dubester, Henry J., The Librarian and the Machine, in *Information Retrieval Today*, ed. Simonton Wesley, University of Minnesota, Center for Continuing Study, Minneapolis, Minn., 1963.
63. Duyvis, Frits Donker, Standardization as a Tool of Scientific Management, *Lib. Trends*, **2**:410-427 (1954).

64. Evans, Richard A., Why and What to Automate, in *Initiating a Library Automation Program,* Special Libraries Association, Documentation Group, Washington, D.C., 1966, pp. 4-8.

65. Fasana, P. J., *Determining the Cost of Library Automation, ALA Bul.,* **61**:656-661 (1967).

66. Ferguson, E., Librarian and the Organization Man, *Sp. Lib.,* **48**:367-372 (1957).

67. Flood, Merrill M., Systems Approach to Library Planning, *Lib. Q.,* **34**:326-338 (1964).

68. *Flowcharting Techniques* (IBM Data Processing Techniques), International Business Machines Corporation, Data Processing Division, White Plains, N. Y., 196?.

69. Freiser, Leonard H., Technology in the Library, *Wilson Lib. Bul.,* **41**:69-71 (1966).

70. Furth, Stephen E., Continuing Education for the Librarian, in *Data Processing in Public and University Libraries,* ed. John Harvey, Spartan Books, Washington, D.C., 1966, pp. 21-23.

71. Gentle, Edgar C. Jr., *Data Communications in Business,* American Telephone and Telegraph Company, New York, 1965.

72. Goldhor, Herbert, Scientific Management in Public Libraries, *Lib. Trends,* **2**:368-389 (1954).

73. Gordon, Paul J., All Very Well in Practice! But How Does It Work Out in Theory?, *Wilson Lib. Bul.,* **42**: 676-685 (1968).

74. Gore, Daniel, A Modest Proposal for Improving the Management of College Libraries, *Educational Record,* **48**:89-96 (1967).

75. Graziano, Eugene E., "Machine-Men" and Librarians, an Essay, *Coll. & Res. Lib.,* **28**:403-406 (1967).

76. Greenwood, James W. Jr., *EDP: The Feasibility Study—Analysis and Improvement of Data Processing* (Systems Education Monograph No. 4), Systems and Procedures Association, Washington, D.C., 1962.

77. Gregory, Robert H. and Richard L. Van Horn, *Automatic Data Processing Systems: Principles and Procedures,* 2nd ed., Wadsworth, Belmont, Cal., 1963.

78. Griffin, H. L., Estimating Data Processing Costs in Libraries, *Coll. & Res. Lib.,* **25**:400-403+ (1964).

79. Grosch, A. N., Special Librarian's Link to Data Processing, *Sp. Lib.,* **57**:635-640 (1966).

80. Grossman, Alvin and Robert L. Howe, *Data Processing for Educators,* Educational Methods, Chicago, 1966.

81. Gull, C. D., Attitudes and Hopes Where Automation Is Concerned in *Automation in the Library—When, Where, and How,* ed. Theodora Andrews with the assistance of Molete Morelock, Purdue University, LaFayette, Indiana, 1965.

82. Gull, C. D., Logical Flow Charts and Other New Techniques for the Administration of Libraries and Information Centers, *Lib. Resources & Tech. Serv.,* **12**:47-66 (1968).

83. Gull, C. D., Personnel Requirements for Automation in Libraries, in *Data Processing in Public and University Libraries,* ed. John Harvey, Spartan Books, Washington, D.C., 1966, pp. 125-141.

84. Haas, Warren J., Computer Simulations at the Columbia University Libraries, in *Clinic on Library Applications of Data Processing, Proceedings, 1964,* ed. Herbert Goldhor, University of Illinois, Graduate School of Library Science, Distributed by the Illini Unon Bookstore, Champaign, Ill., 1965, pp. 36-46.

85. Hage, Elizabeth B., An Administrator's Approach to Automation at the Prince George's County (Maryland) Memorial Library, in *Clinic on Library Applications of Data Processing, Proceedings, 1967,* ed. Dewey E. Carroll, University of Illinois, Graduate School of Library Science, Urbana, Ill., 1967, pp. 90-97.

86. Hammer, Donald P., Scheduling Conversion, in *Data Processing in Public and University Libraries,* ed. John Harvey, Spartan Books, Washington, D.C., 1966, pp. 103-123.

87. Hardkopf, J. C. M., Cybernetics and the Library, *Lib. J.,* **76**:999-1001 (1951).

88. Hardkopf, J. C. M., Major Book Move; Time Study, *Lib. J.,* **80**:2417-2419 (1955).

89. Hausdorfer, W, Guidance for Administration, *Lib. Trends,* **7**:481-491 (1959).

90. Hayes, Robert M., *The Concept of an On-line Total Library System,* (ALA Library Reports), American Library Association, Chicago, Ill. 1965.

91. Hayes, Robert M., Data Processing in the Library School Curriculum *ALA Bul.,* **61**:662-668 (1967).

92. Hayes, Robert M., Implications for Librarianship of Computer Technology, in *Clinic on Library Applications of Data Processing, Proceedings, 1964,* ed. Herbert Goldhor, University of Illinois, Graduate School of Library Science, Distributed by the Illini Union Bookstore, Champaign, Ill., 1965, pp. 1-6.

93. Hayes, Robert M., Library Systems Analysis, in *Data Processing in Public and University Libraries,* ed. John Harvey, Spartan Books, Washington, D.C., 1966, pp. 5-20.

94. Hayes, Robert M., The Meaning of Automation to the Library Profession, *PNLA Q.,* **27**:7-16 (Oct. 1962).

95. Heiliger, E. M., Staffing a Computer-based Library, *Lib. J.,* **89**:2738-2739 (1964).

96. Herner, Saul, Meaningful Statistics, in *Practical Problems of Library Automation,* Special Libraries Association, Documentation Group, Washington, D.C., 1967, pp. 47-52.

97. Herner Saul, System Design, Evaluation, and Costing, *Sp. Lib.,* **58**:576-581 (1967).

98. Hicks, Charles B. and Irene Place, *Office Management,* 2nd ed., Prentice-Hall, Englewood Cliffs, N. J., 1962.

99. Hines, Theodore C., Computers, Supervisors, Libraries, *ALA Bul.* **62**:153-157 (1968).

100. Howard, Paul, Consequences of Management Surveys, *Lib. Trends,* **2**:428-436 (1954).

101. Howe, Mary T., The Establishment and Growth of the Data Processing De-

partment in the Decatur Public Library, in *Data Processing in Public and University Libraries,* ed. John Harvey, Spartan Books, Washington, D.C., 1966, pp. 37-52.

102. Johnson, Richard A., *et al, The Theory and Management Systems,* McGraw-Hill, New York, 1963.

103. *Initiating a Library Automation Program,* Special Libraries Association, Documentation Group, Washington, D.C., 1966.

104. Jackson, Ivan F., An Approach to Library Automation Problems, *Coll. & Res. Lib.,* **28**:133-137 (1967).

105. Jestes, Edward C., An Example of Systems Analysis: Locating a Book in a Reference Room, *Sp. Lib.,* **59**:722-728 (1968).

106. Kapp, R. O., The First Draft, in *Computer Peripherals and Typesetting,* by Arthur H. Phillips, H.M.S.O., London, 1968, pp. 603-611.

107. Keller, John E., Program Budgeting and Cost Benefit Analysis in Libraries, *Coll. & Res. Lib.,* **30**:156-160 (1969).

108. Kennedy, M., Management Methods in Libraries; a Symposium. Procedure Analysis, *Med. Lib. Assn. Bul.,* **49**:520-522 (1962).

109. Kent, A., *Library Planning for Automation,* Spartan Books, Washington, D.C., 1965.

110. Keys, T. E., Changing Concepts in Library Services, *Med. Lib. Assn. Bul.,* **45**:5-13 (1957)

111. Kilgour, Frederick G., Basic Systems Assumptions of the Columbia-Harvard-Yale Medical Libraries Computerization Project, in *Institute on Information Storage and Retrieval, 1965,* University of Minnesota, Minneapolis, Minn., 1966, pp. 145-154.

112. Kilgour, Frederick G., Comprehensive Modern Library Systems, in *Brasenose Conference on the Automation of Libraries,* ed. John Harrison and Peter Laslett, Mansell, London, 1967.

113. Kilgour, Frederick G., Systems Concepts and Libraries, *Coll. & Res. Lib.* **28**:167-170 (1967).

114. Kimber, Richard T., *Automation in Libraries* (International Series of Monographs in Library and Information Science v. 10), Pergamon, Oxford, England, 1968.

115. Kipp, Laurence J., Scientific Management in Research Libraries, *Lib. Trends,* **2**:390-400 (1954).

116. Knox, W. T., Changing Role of Libraries, *ALA Bul.,* **59**:720-725 (1965).

117. Koriagin, Gretchen W., Experience in Man and Machine Relationships in Library Mechanization, *Am. Doc.,* **15**:227-229 (1964).

118. Kountz, John C., Cost Comparison of Computer versus Manual Catalog Maintenance, *J. Lib. Automation,* **1**:159-177 (1968).

119. Kozumplik, William A., Time and Motion Study of Library Operations, *Sp. Lib.,* **58**:585-588 (1967).

120. Kraft, Donald R., Basic Computer Information for Libraries, in *Automation in the Library—When, Where, and How,* ed. Theodora Andrews with the as-

sistance of Molete Morelock, Purdue University, LaFayette, Ind., 1965, pp. 3-22.

121. Krikelas, J., Library Statistics and the Measurement of Library Services, *ALA Bul.,* **60**:494-499 (1966).

122. Kurmey, William J., Management Implications of Mechanization, in *Automation in Librares,* Canadian Association of College and University Libraries, University Libraries, University of British Columbia, Vancouver, 1967, pp. 116-123.

123. Laden, H. N. and T. R. Gildersleeve, *System Design for Computer Applications,* John Wiley, New York, 1963.

124. Lamkin, Burton E., Decision-making Tools for Improved Library Operations, *Sp. Lib.,* **56**:642-646 (1965).

125. Lamkin, Burton E., Systems Analysis in Top Management Communication, *Sp. Lib.,* **58**:90-94 (1967).

126. Lazorick, G. J. and T. L. Minder, Least Cost Searching Sequence, *Coll. & Res. Lib.,* **25**:126-128 (1964)

127. Lebowitz, Abraham I., The Mechanization of Technical Services in Special Libraries, in *Practical Problems of Library Automation,* Special Libraries Association, Documentation Group, Washington, D.C., 1967, pp. 1-6.

128. Lee, Chong Hak, *The Impact of EDP Upon the Patterns of Business Organizations and Administration,* State University of New York, School of Business, Albany, N. Y., 1965.

129. Leimkuhler, F. F. and J. G. Cox, Compact Book Storage in Libraries, *J. Oper. Res. Soc of America,* **12**:419-427 (1967)

130. Leimkuhler, F. F., Operations Research in the Purdue Libraries, in *Automation in the Library—When, Where, and How,* ed. Theodora Andrews with the assistance of Molete Morelock, Purdue University, Lafayette, Ind., 1965, pp. 82-89.

131. Leimkuhler, Ferdinand F., Systems Analysis in University Libraries, *Coll. & Res. Lib.,* **27**:13-18 (1966).

132. Levy, Joseph, *Punched Card Data Processing,* McGraw-Hall, New York, 1967.

133. Line, Maurice Bernard, *Library Survey,* Archon Books, Hamden, Conn., 1967.

134. Littlefield, C. L. and Frank Rachel, *Office and Administrative Management,* 2nd ed., Prentice-Hall, Englewood Cliffs, N. J., 1964.

135. Logsdon, Richard H., Time and Motion Studies in Libraries, *Lib. Trends,* **2**:401-409 (1954).

136. Lott, Richard W., *Basic Data Processing,* Prentice-Hall, Englewood Cliffs, N. J., 1967.

137. Lynch, M. F., The Library and the Computer, in *The Library and the Machine,* ed. C. D. Batty, North Midland Branch of the Library Association, obtainable from the Public Library, Scunthorpe (Lincs.), 1966, pp. 21-36.

138. McDiarmid, Errett W., Scientific Method and Library Administration, *Lib. Trends,* **2**:361-367 (1954).

139. MacKenzie, A. Graham, Systems Analysis of a University Library, *Program,* **2**:7-14 (April 1968).

140. Markuson, B. E., ed., *Libraries and Automation,* Library of Congress, Washington, D.C., 1964.

141. Martin, E. Wainright, *Electronic Data Processing; an Introduction,* Rev. ed., Irwin, Homewood, Ill., 1965.

142. Meacham, Alan D. and Van B. Thompson, ed., *Total Systems,* American Data Processing Inc., Detroit, 1962.

143. Meise, Norman R., *Conceptual Design of an Automated National Library System* (Master's Thesis), Rensselaer Polytechnic Institute, Hartford Graduate Center, Hartford, Conn, 1966.

144. Melin, John S., *Libraries and Data Processing—Where Do We Stand?,* (University of Illinois, Graduate School of Library Science Occasional Paper, No. 72), University of Illinois, Graduate School of Library Science, Urbana, Ill., 1964.

145. Memo on Effective Labor Costs, in *Case Studies in Systems Analysis in a University Library,* ed. Barton R. Burkhalter, Scarecrow Press, Metuchen, N. J., 1968, pp. 9-10.

146 Minder, T. L., Library Systems Analyst: a Job Description, *Coll. & Res. Lib.,* **27**:271-276 (1966).

147. Minder, Thomas, The Problem Solving Routine, in *Initiating a Library Automation Program,* Special Libraries Association, Documentation Group, Washington, D.C., 1966, pp. 1-3.

148. Moore, Edythe, Systems Analysis: An Overview, *Sp. Lib.,* **58**:87-90 (1967).

149. Morsch, Lucile M., Scientific Management in Cataloging, *Lib Trends,* **2**:470-483 (1954).

150. Mullee, William Robert, *Better Communications and Control through Records and Reports Study,* Work Simplification Round Tables, Loyola University, Los Angeles, Cal., 1963.

151. Mullee, William Robert, *The Flow Process Chart,* Work Simplification Round Tables, Loyola University, Los Angeles, Cal., 1958.

152. Mundel, M. E., *Motion and Time Study; Principles and Practice,* 3rd ed., Prentice-Hall, Englewood Cliffs, N. J., 1960.

153. Myatt, D. O. and J. Entel, Profit Viewpoint in Library Management and Operation, *Sp. Lib.,* **48**:373-377 (1957).

154. Nance, Richard E., Systems Analysis and the Study of Information Systems, *American Documentation Institute. Proceedings,* **4**:70-74 (1967).

155. The Needs of Libraries, in *EDUNET; Report of the Summer Study on Information Networks . . . ,* John Wiley, New York, 1967, pp. 61-71.

156. Neumaier, Richard, comp., *Better Business Forms,* the author, Philadelphia, Pa., 1963.

157. Neuschel, Richard F., *Management by System,* McGraw-Hill, New York, 1960.

158. Newell, W. T., Long-range Planning for Library Managers, *PNLA Q.,* **31**:21-35 (1966).

159. Nicolaus, John J., How to Begin: Initiating a Library Automation Program, in *Initiating a Library Automation Program,* Special Libraries Association, Documentation Group, Washington, D.C., 1966, pp. 9-23.

160. Oh, T. K., New Dimensions of Management Theory, *Coll. & Res. Lib.*, **27**:431-438 (1966).

161. *Online Library Circulation Control System, Moffet Library Midwestern University, Wichita Falls, Texas* (IBM Application Brief), International Business Machines Corporation, Data Processing Division, White Plains, N. Y., 1969.

162. *Operations Research; Challenge to Modern Management,* Harvard University, Graduate School of Business Administration, Cambridge, Mass., 1954.

163. Optner, Stanford L., *Systems Analysis for Business Management,* 2nd ed., Prentice-Hall, Englewood Cliffs, N. J., 1968.

164. Ottemiller, John H., The Management Engineer, *Lib. Trends,* **2**:437-451 (1954).

165. Parker, R. H., Basic Concept of Data-Processing for Libraries, in *Proceedings of the National Conference on the Implications of the New Media for the Teaching of Library Science,* ed. Harold Goldstein, University of Illinois, Chicago, Ill., 1963.

166. Parker, Ralph H., Concept and Scope of Total Systems in Library Records, in *Data Processing in Public and University Libraries,* ed. John Harvey, Spartan Books, Washington, D.C., 1966, pp. 67-77.

167. Parker, Ralph H., Development of Automatic Systems at the University of Missouri Library, in *Clinic on Library Applications of Data Processing, Proceedings, 1963,* ed. Herbert Goldhor, University of Illinois, Graduate School of Library Science, distributed by the Illini Union Bookstore, Champaign, Ill., pp. 43-55.

168. Parker, Ralph H., Economic Consideration, in *Data Processing in Public and University Libraries,* ed. John Harvey, Spartan Books, Washington, D.C., 1966, pp. 143-147.

169. Parker, Ralph H., Library Records in a Total System, in *The Brasenose Conference on the Automation of Libraries,* ed. John Harrison and Peter Laslett, Mansell, London, 1967.

170. Parker, Ralph H., The Machine and the Librarian, *Lib. Resources & Tech. Serv.,* **9**:100-103 (1965).

171. Parker, Ralph H., Not a Shared System, *Lib. J.,* **92**:3967-3970 (1967).

172. Parker, Ralph H., What Every Librarian Should Know about Automation, *Wilson Lib. Bul.,* **38**:752-754 (1964).

173. Pflug, Gunther, Problems of Electronics Data Processing in Libraries, *Libri,* **15**:35-49 (1965).

174. Poage, S. T., Work Sampling in Library Administration, *Lib. Q.,* **30**:213-218 (1960).

175. Pings, Vern M., Development of Quantitative Assessment of Medical Libraries, *Coll. & Res. Lib.,* **29**:373-380 (1968).

176. *Practical Problems of Library Automation,* Special Libraries Association, Documentation Group, Washington, D.C., 1967.

177. Quatman, G. L., *Costs of Provding Library Services to Groups in the Purdue University Community,* Purdue University Libraries, Lafayette, Ind., 1962.

178. Raphael, David L., *Analyzing Staff Utilization in an Academic Library* (presented at the 86th Annual Conference of American Library Association, San Francisco, June 27, 1967), mimeo.

179. *Research Methods in Librarianship: Measurement and Evaluation,* (University of Illinois, Graduate School of Library Science Monograph Series Monograph Series No. 8), University of Illinois, Graduate School of Library Science, Champaign, Ill., 1968.

180. Risk, J. M. S., Information Services: Measuring the Cost, *Aslib Proc.,* **8**:269-287 (1956).

181. Robichaud, Beryl, *Understanding Modern Business Data Processing,* McGraw-Hill, New York, 1966.

182. Rogers, Frank B., Management Improvement in the Library, *Med. Lib. Assn. Bul.,* **48**:404-409 (1957).

183. St. John, F. R., Management Improvements in Libraries, *Coll. & Res. Lib.,* **14**:174-177 (1953).

184. Salvail, J. A., Management's Prescription for 'Computeritis,' *Software Age,* **2**:16-20 (May 1968).

185. Schmid, Calvin F., *Handbook of Graphic Presentation,* Ronald Press, New York, 1954.

186. Schuler, S., Where Can You Cut Costs Next?, *Nation's Business,* **51**:58-60+ (Oct. 1963).

187. Schultheiss, Louis A., *et al, Advanced Data Processing in the University Library,* Scarecrow Press, New York, 1962.

188. Schultheiss, Louis A., Systems Analysis and Planning, in *Data Processing in Public and University Libraries,* ed. John Harvey, Spartan Books, Washington, D.C., 1966, pp. 95-102.

189. Schultheiss, Louis A., Automation of Library Operations, in *Computer Applications Symposium, Proceedings,* MacMillan, New York, 1962, pp. 35-44.

190. Schultheiss, Louis A., Techniques of Flow-Charting, in *Clinic on Library Applications of Data Processing. Proceedings,* 1963, ed. Herbert Goldhor, University of Illinois, Graduate School of Library Science, Distributed by the Illini Union Bookstore, Champaign, Ill., 1964, pp. 62-78.

191. Setting Work Standards, *Administrative Management,* **24**:65-67 (Sept. 1963).

192. Shaw, Ralph R., Management, Machines and the Bibliographic Problems of the Twentieth Century, in *Bibliographic Organization,* ed. Jesse H. Shera and Margaret E. Egan, University of Chicago, Chicago, Ill., 1951, pp. 200-225.

193. Shaw, Ralph R., Scientific Management in Libraries, *Lib. Trends,* **2**:359-360 (1954).

194. Shera, J. H., Automation without Fear, *ALA Bul.,* **55**:787-794 (1961).

195. Shera, Jesse H., Librarians against Machines, *Wilson Lib. Bul.,* **42**:65-73 (1967).

196. Shera, Jesse H., Librarian and the Machine, *Lib. J.,* **86**:2250-2254 (1961).

197. Sigband, Norman B., *Effective Report Writing for Business, Industry, and Government,* Harper and Row, New York, 1960.

198. Simms, Daniel M., What Is a Systems Analyst?. *Sp. Lib.,* **59**:718-721 (1968).

199. Sippl, Charles J., *Computer Dictionary and Handbook,* Howard Sams, Indianapolis, Ind., 1966.

200. Smith, Paul T., *How to Live with Your Computer,* American Management Association, New York, 1956.

201. Solomon, Irving I. and Laurence O. Weingart, *Management Uses of the Computer,* Harper and Row, New York, 1966.

202. *Standard Manual of Paperwork Flow Charting for Management,* Capstone Book Press, Arvada, Col., 1967.

203. Stein, Theodore, Automation and Library Systems, *Lib. J.,* **89**:2723-2734 (1964).

204. Stewart, Bruce W., Data Processing in an Academic Library, *Wilson Lib. Bul.,* **41**:388-395 (1966).

205. Strain, Paula M., Second Thoughts on Serials Automation, *SLA Sci-Tech News,* **23**:9-10 (Spring, 1969).

206. Stuart-Stubbs, Basil, Automation in a University Library from the Administrator's Viewpoint, in *Automation in Libraries,* Canadian Association of College and University Libraries, University of British Columbia, Vancouver, Canada, 1967, pp. 124-134.

207. Sullivan, John W., Increasing Computer Efficiency: Educate All Personnel, *Advanced Management J.,* **29**:17-19 (1964).

208. Swanson, Don R., Library Goals and the Role of Automation, *Sp. Lib.* **53**:466-471 (1962).

209. Swanson, Don R., *Library Service with or without Automation* (Occasional Paper No. 61), Canadian Library Association, Ottawa, Canada, 1965.

210. Systems and Procedures Association, *Business Systems,* The Association, Cleveland, Ohio, 1966.

211. Tauber, M. F., Survey Methods in Approaching Library Problems, *Lib. Trends,* **13**:15-30 (1964).

212. Taylor, Robert S., An Approach to Library Systems Analysis (Library Systems Analysis Report No. 2), Lehigh University, Center for the Information Sciences, Bethlehem, Pa., 1964.

213. Taylor, Robert S. and Caroline E. Hieber, *Library Systems Analysis* (Library Systems Analysis Report No. 3), Lehigh University, Center for the Information Sciences, Bethlehem, Pa., 1965.

214. Taylor, Robert S., Question-Negotiation and Information Seeking in Libraries, *Coll. & Res. Lib.,* **29**:178-194 (1968).

215. Thompson, Victor A., The Organizational Dimension, *Wilson Lib. Bul.,* **42**:693-700 (1968).

216. Tichy, Henrietta J., *Effective Writing for Engineers, Managers, Scientists,* John Wiley, New York, 1966.

217. Tompkins, Marjorie M., Classification Evaluation of Professional Librarian Positions in the Universtiy of Michigan Library, *Coll. & Res. Lib.,* **27**:175-184 (1966).

218. Tosi, Henry L., Admnistrators' Development Program, *Wilson Lib. Bul.,* **42**:406-410 (1967).

219. Trueswell, Richard W., A Quantitative Measure of User Circulation Require-

ments and Its Possible Effect on Stack Thinning and Multiple Copy Determination, *American Documentation,* **16**:20-25 (1965).

220. Trueswell, Richard W., Some Behavioral Patterns of Library Users: the 80/20 Rule, *Wilson Lib. Bul.,* **43**:458-461 (1969).

221. Trueswell, Richard W., Some Circulation Data from a Research Library, *Coll. & Res. Lib.,* **29**:493-495 (1968).

222. Trueswell, Richard W., Two Characteristics of Circulation and Their Effect on the Implementation of Mechanized Circulation Control Systems, *Coll. & Res. Lib.,* **25**:285-291 (1964).

223. Ulman, J. N., Jr. and J. R. Gould, *Technical Reporting,* Rev. ed., Holt, New York, 1959

224. Van Pelt, J. D., Time and Motion Study in a Catalogue Room, *Australian Lib. J.,* **5**:55-63 (1956).

225. *Voos, Henry, Standard Times for Certain Clerical Activities,* in Technical Processing (Doctoral Thesis), Rutgers University, New Brunswick, N. J., 1965.

226. Wainwright, Martin E., *Electronic Data Processing,* Rev. ed., Irwin, Homewood, Ill., 1965.

227. Waite, David P., Developing a Library Automation Program, *Wilson Lib. Bul.,* **43**:52-58 (1968).

228. Wallace, W. Lyle, ed., *Work Simplification* (Systems Education Monograph No. 1), Systems and Procedures Association, Detroit, Mich., 1962.

229. Wasserman, Paul, *The Librarian and the Machine,* Gale, Detroit, Mich., 1965.

230. Wasserman, Paul, Policy Formulation in Libraries, *Ill. Lib.,* **43**:772-779 (1961).

231. Watson, William, Library Automation: a Primer on Some of the Implications, in *Automation in Libraries,* Canadian Association of College and University Libraries, University of British Columbia, Vancouver, Canada, 1967, pp. 135-146.

232. Weed, K. K., Tool for Management Evaluation of Library Services, *Sp. Lib.,* **48**:378-382 (1957).

233. Weindling, Ralph, *et al, A Management Guide to Computer Feasibility,* American Data Processing, Inc., Detroit, Mich., 1962.

234. Welsh, William J., Compatibility of Systems, in *Data Processing in Public and University Libraries,* ed. John Harvey, Spartan Books, Washington, D.C., 1966, pp. 79-93.

235. Wertz, John A., Possible Application of Data Processing Equipment in Libraries, in *Clinic on Library Applications of Data Processing. Proceedings,* **1964**, ed. Herbert Goldhor, University of Illinois, Graduate School of Library Science, Distributed by the Illini Union Bookstore, Champaign, Ill., 1965, pp. 112-117.

236. Wheeler, Gershon J., *Business Data Processing: an Introduction,* Addison-Wesley, Reading, Mass., 1966.

237. Williams, Lawrence K., Managing Change: a Test of the Administration, *Wilson Lib. Bul.,* **42**:686-692 (1968).

238. Withington, Frederic G., *The Use of Computers in Business Organizations,* Addison-Wesley, Reading, Mass., 1966.

239. Woodruff, E. L. Work Measurement Applied to Libraries, *Sp. Lib.,* **48**:139-144 (1957).

240. White, Herbert S., Where and When to Begin, in *Initiating a Library Automation Program* Special Libraries Association, Documentation Group, Washington, D.C., 1966, pp. 24-35.

241. Wyld, Lionel D., *Preparing Effective Reports,* Odyssey, New York 1967.

242. Wylie, Harry L., *Office Management Handbook,* 2nd ed., Ronald Press, New York, 1958.

243. Wynar, Don, *et al,* Cost Analysis in a Technical Services Division, *Lib. Resources & Tech. Serv.,* **7**:312-326 (1963).

BIBLIOGRAPHY OF BIBLIOGRAPHIES

1. Henderson, Madeline M., *Evaluation of Information Systems: A Selected Bibliography with Informative Abstracts* (National Bureau of Standards, Technical Note No. 297), U.S.G.P.O., Washington, D.C., 1967.

2. McCune, L. C. and S. R. Salmon, Bibliography of Library Automation, *ALA Bul.,* **61**:674-675+ (1967).

3. Markuson, Barbara Evans, comp., Automation in Libraries and Information Centers, in *Annual Review of Information Science and Technology* (Vol. 2), John Wiley (Interscience), New York, 1967, pp. 255-284.

4. *New Publications in the Unesco Library, Selected Bibliography on Computers, Data Processing and Allied Subjects, 1960-67,* Unesco Library, Paris, June, 1968.

5. Speer, Jack A., comp., *Libraries and Automation, a Bibliographly with Index,* Teachers College Press, Kansas State Teachers College, Emporia, Kan., 1967.

6. Taylor, Robert S., *Bibliography of Recent Reports and Books on Library Automation and Library Systems Analysis* (Library Systems Analysis Report No. 1), Lehigh University, Center for the Information Sciences, Bethlehem, Pa., 1964.

7. Wessel, C. J., and B. A. Cohrssen, *Criteria for Evaluating the Effectiveness of Library Operations and Services. Phase I: Literature Search & State of the Art* (ATLIS Report No. 10), John Thompson & Co., Washington, D.C., 1967, available from clearinghouse, AD649468.

INDEX